P9-DWH-915

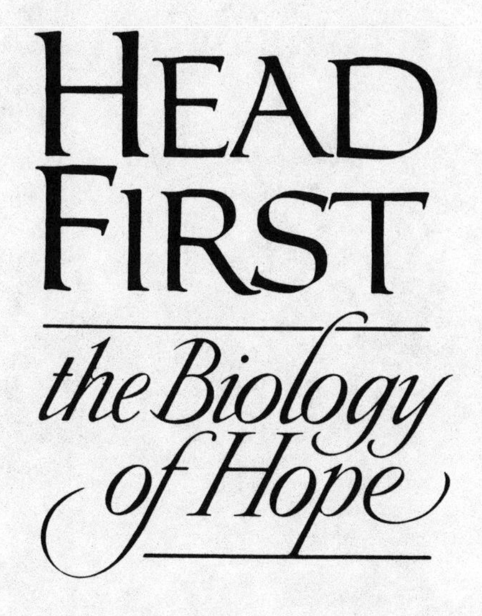
HEAD
FIRST
the Biology
of Hope

BOOKS BY NORMAN COUSINS

The Good Inheritance: The Democratic Chance

Modern Man is Obsolete

Talks with Nehru

Who Speaks for Man?

Dr. Schweitzer of Lambarene

In Place of Folly

Present Tense: An American Editor's Odyssey

The Improbable Triumvirate: An Asterisk to the History of a Hopeful Year 1962–63

The Celebration of Life: A Dialogue on Immortality and Infinity

Anatomy of an Illness as Perceived by the Patient: Reflections on Healing and Regeneration

Human Options: An Autobiographical Notebook

The Healing Heart: Antidotes to Panic and Helplessness

Albert Schweitzer's Mission: Healing and Peace

The Human Adventure: A Camera Chronicle

The Pathology of Power

BOOKS EDITED BY NORMAN COUSINS

A Treasure of Democracy

The Poetry of Freedom

Writing for Love or Money

March's Thesaurus-Dictionary

Great American Essays

Profiles of Gandhi

Memoirs of a Man—Grenville Clark

The Physician in Literature

The Republic of Reason: The Personal Philosophies of the Founding Fathers

K. Jason Sitewell's Book of Spoofs

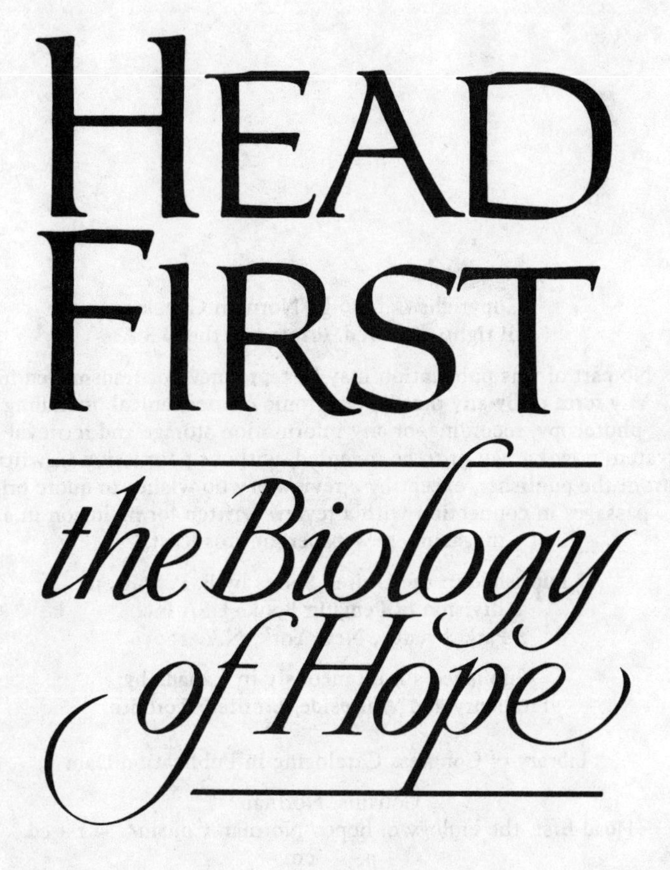

NORMAN
COUSINS

E. P. DUTTON  NEW YORK

Published in the United States by E. P. Dutton,
a division of Penguin Books USA Inc.,
2 Park Avenue, New York, N.Y. 10016.

Published simultaneously in Canada by
Fitzhenry and Whiteside, Limited, Toronto.

Library of Congress Cataloging-in-Publication Data

Cousins, Norman.
Head first, the biology of hope / Norman Cousins. — 1st ed.
p. cm.
Includes index.
ISBN 0-525-24805-6
1. Medicine, Psychosomatic. I. Title.
RC49.C674 1989
616.08—dc20 89-7789
CIP

DESIGNED BY EARL TIDWELL

DEDICATION

DEAN SHERMAN MELLINKOFF
who helped me to convert curiosity into scientific pursuit

and

JOAN B. KROC
who lives out her dream to put
the dreams of others within reach.

Just before his death, Socrates remembered a debt. "I owe a cock to Asklepios," he said, referring to the Greek custom of propitiating the god of medicine following medical treatment.

The Latinized word for Asklepios is Aesculapius. Those who commit their lives to service in medicine are called Aesculapians. This volume acknowledges my ten years among the Aesculapians.

CONTENTS

AUTHOR'S NOTES

The word *acknowledgment* is much too pallid and passive to describe my debt to the persons who helped in this book. Indeed, it is important to emphasize that this book is a team effort. I was the beneficiary of essential advice and assistance in all the phases of book preparation and production—research, references, fact-checking, copy-editing—and in the viewpoints expressed in this volume.

At the UCLA School of Medicine, Dean Sherman M. Mellinkoff and his successor, Dr. Kenneth I. Shine; Dr. L. Jolyon West, Chairman of the Department of Psychiatry and Biobehavioral Sciences; Dr. Carmine Clemente, Professor of Anatomy and former Director of the Brain Research Institute at UCLA; Dr. Mitchel D. Covel, Assistant Dean; and Dr. George Solomon, Adjunct Professor of Psychiatry, University of California at San Francisco all provided valuable criticisms. Jean Anderson, in addition to her editorial assistance, presided over the word processor and produced the dozen or more revisions and drafts leading to the final manuscript. Drawing upon her work as coordinator of the UCLA Task Force in Psychoneuroimmunology, Ping Ho assembled the studies described in this book and superintended the glossary, honor roll, and references. Janet Thomas ordered and made accessible the many research materials required. Dr. Evelyn Silvers and Dori Lewis did painstaking line-by-line editing and made substantive suggestions. Julian Bach, lifelong friend, put his editorial expertise at my disposal in helping to define the purposes and audience of this book.

A separate and special paragraph is owed to Joyce Engelson, editor-in-chief of E. P. Dutton, who applied uncommon editorial skills to the making of the book and who exercised supernatural patience in shepherding the author through a succession of drafts.

To all these collaborators, I make known my deep appreciation. Obviously, any flaws, failures, errors, or inadequacies are chargeable solely to the author.

The dedication to Joan B. Kroc is a reflection of her role in establishing the Program in Psychoneuroimmunology at the UCLA School of Medicine.

Road Map: In order to avoid encumbering the text with footnotes, a special section in the back of the book identifies, chapter by chapter, the sources of reports, papers, articles, books, and other references cited in the text.

The medical terminology used in the book is, I believe, generally understandable in the context in which it appears, but a guide and explanation of technical terms will be found at the back of the book. Many readers will, I know, find such scientific terminology helpful and convincing, but the points the book makes are understandable without the medical or academic backup.

This book does not purport to be a history of mind-body theory and research. However, in the chapter called "Portfolio of Related Matters," an "Honor Roll" seeks to identify a number of medical investigators and thinkers over the centuries who have recognized the interactive workings of body and mind. The roster of contemporary names in the Honor Roll, admittedly, is incomplete. Research in the general field of psychoneuroimmunology turns up new major contributors each month. The omission of any names, whether in the body of the book or the Honor Roll, is not to be construed as judgment of their work.

HEAD
FIRST
the Biology
of Hope

1

BIRTH OF AN OBSESSION

The scene is a research laboratory in the Health Sciences complex at the University of California at Los Angeles. Seated in a chair is a professional actor. She is listening carefully to instructions from a medical researcher.

"Imagine that you are in your first Broadway play," the researcher is saying. "You have the leading role. All your years of training, all your hopes, have pointed to this moment and you have poured all your talent and energy into this performance. The success of everything you and your fellow actors have done depends largely on the review by the drama critic of *The New York Times*. After the play, you join the producer, the director, and the other actors at Sardi's restaurant to wait for the review.

"The wait seems like a lifetime. Then someone rushes into the restaurant waving the early-morning edition of *The Times*. The producer grabs the newspaper, flashes to the drama review, and begins to read aloud. It is a smash hit and the drama critic predicts a long run. He singles out your performance as one of the best he has ever seen. Now, how do you feel? What do you do? What do you say? Act now as you would if you were in Sardi's and were in the center of all this wonderful attention."

The actor, a young woman, sits back deep in her chair. She puts both hands to her face. Eyes and mouth are wide open. Then, after a second or two, comes the eruption. She leaps out of her seat as though propelled by rocket fuel.

"Whee!" she cries. It is the scream of elation so readily recognized by Broadway first-nighters who wait up at Sardi's for the first reviews.

"I can't believe it, I can't believe it!" she exclaims. "How wonderful for all of us."

Unable to contain her joy, she pantomimes embracing and kissing her companions.

A blood sample is drawn from her arm. Blood pressure measurements are taken. So are galvanic skin changes and other vital signs.

The purpose of the experiment is to see whether the positive emotions—in particular, happiness—can have positive physiological effects, including an enhanced immune system.

This particular experiment was undertaken in conjunction with UCLA's Program in Psychoneuroimmunology, a new branch of medicine based on the interaction of the brain, the endocrine system, and the immune system, and the task force at UCLA carrying out this program is one of the few such groups of its kind. It was formed for the express purpose of investigating the way the emotions make their registrations on the body. One of the possible applications of the study was the role of attitudes in combating serious illness.

In a sense, the formation of the task force and the research project involving the physiological effects of the emotions of actors were both connected to the obsession that brought me to UCLA a decade earlier. This obsession was the need to find proof, or at least to help create it, that the human brain could bring about changes for the better in the way human beings confront illness. There was abundant medical research to show that the brain, under circumstances of the negative emotions—hate, fear, panic, rage, despair, depression, exasperation, frustration—could produce powerful changes in the body's chemistry, even set the stage for intensified illness. But there was no comparable evidence to show that the positive emotions—purpose, determination, love, hope, faith, will to live, festivity—could also affect biological states. It is well known that we have the ability to make ourselves ill. What about the ability to make ourselves well? Suppose it could be proved that positive attitudes might actually help activate healing forces in the endocrine and immune systems? What would the implications of such findings be on the treatment of serious illness?

There was nothing original about these questions, of course. Hippocrates, father of medicine, insisted that medical students give full weight to the emotions, both as a contributing cause of disease and as a factor in recovery. Aristotle discoursed at length about the role of the emotions in health and illness. Throughout history, physicians emphasized the importance of the patient's will to live in treating disease. What was missing, however, was precise information about the biological effects of the positive emotions. If emotions such as hope and purpose actually produced physiological change, exactly what mechanisms and pathways were involved?

I was warned that the thesis or assumption underlying this quest would not meet with universal enthusiasm within the medical community.

"Physicians regard themselves as scientists and most of them believe that nothing is more unscientific than the notion that willpower or attitudes have anything to do with overcoming serious disease," Dr. Franklin Murphy told me. Dr. Murphy, a physician himself, was formerly chancellor of the University of California at Los Angeles.

"I've spoken to doctors who have read your articles on these matters," he continued. "Frankly, some of them think you're way off base. Fortunately, there are other doctors who have seen cases in their own practice—patients who have survived serious illness not just because of medical treatment but because of a powerful will to live. Those doctors may not be in the majority—at least not yet—but you will not be without allies in the belief that emotions and attitudes can affect the body's chemistry."

The prospect of allies within the medical community was encouraging. But I'm getting a little ahead of my story. Until 1978, my entire professional life had been spent in writing and editing, with brief interludes in teaching and government service. I had a deep interest in medical science and the human healing system, reflected in special sections appearing from time to time in the *Saturday Review*, where I had been editor from

1940 to 1971. I also had the benefit of my own body as a laboratory, having recovered fully from severe personal illnesses. Franz Ingelfinger, the editor of the *New England Journal of Medicine,* had encouraged me to contribute an article to his periodical about one of these illnesses. That article, perhaps more than any other single factor, was to result in invitations to meet with students and faculty at various medical centers. It was also to lead to the obsession that brought me into the medical community.

Dr. Murphy, with whom I had been associated as fellow trustee on various boards, steered me in the direction of UCLA. He said that if I was determined to pursue my obsession, a good place to do it would be at UCLA. He spoke of the broad research capabilities of UCLA's School of Medicine as well as its interest in humanistic medical education. He described the remarkable growth in size and stature of UCLA's medical school in the past two decades. Its national standing had climbed year by year until it was ranked among the top six medical schools in the country. Much of this recognition, Dr. Murphy said, had come about under Dean Sherman Mellinkoff.

On a parallel track, Dr. Omar Fareed of Los Angeles, whom I had met at the Albert Schweitzer Hospital in Lambaréné in what was then French Equatorial Africa (now West Gabon) and who had been associated with medical projects abroad organized by the *Saturday Review,* had been urging me to explore the possibilities of a connection with UCLA. When we discussed an invitation of the Columbia College of Physicians and Surgeons in New York, Omar felt I ought to pursue Dr. Murphy's advice about the advantages of UCLA.

Yet another point of possible entry was furnished by Dr. Bernard Towers, the distinguished English physician-philosopher who had come to UCLA several years earlier and who had initiated a forum at the medical school under the title Society and Human Values. My first visit to UCLA came early in 1978 and was in response to an invitation from Dr. Towers to speak at his forum. The subject was the efficacy of the placebo, on which I had recently written for the *Saturday Review.* Dean Mellinkoff would be one of the participants. It was a fortuitous invitation and I was pleased to accept.

During the course of my visit to the medical school I had

an opportunity to spend some time with Dr. Mellinkoff, who had been at UCLA for a quarter of a century. He had been trained as a gastroenterologist and had taught at the Johns Hopkins medical school. His abilities not just as a physician and educator but as a philosopher of medicine won him increasing recognition. He was recruited by UCLA as assistant professor of medicine, becoming dean after the retirement of Dr. Stafford Warren.

Within minutes of arriving at Dr. Mellinkoff's office, I could understand Dr. Murphy's enthusiasm about the man. It was quickly apparent that Dr. Mellinkoff was in the great tradition of philosopher-teacher-practitioner associated with Sir William Osler. Dr. Mellinkoff was interested not just in disease but health, not just in ways of treating illness but in preventing it, not just in identifying harmful microorganisms but in knowing something about the circumstances in which they take hold, and finally in developing the kind of relationships with patients that would give them confidence in themselves and in their physicians.

Dr. Mellinkoff escorted me around the medical school and, indeed, around the university campus. I was aware that UCLA had a growing reputation as a world crossroads of scholarship and culture. Students and teachers from every continent formed a significant part of the university population. I had the feeling that UCLA, with its international flavor and cultural ferment, was not far from being the kind of educational center associated with Heidelberg University at the turn of the century. The physical expanse of the medical school alone was staggering, with more than twenty miles of corridors winding through its various departments, research centers, and hospitals. These dimensions hit home when I learned that the UCLA Center for the Health Sciences occupied more space than any building west of the Pentagon.

At that first meeting the dean and I discussed the need for producing not just well-trained but well-educated physicians. We also discussed our mutual interest in the way physicians are perceived in the world of literature and what would go into a course on the subject. Dr. Mellinkoff said he was eager to have me meet other members of the medical faculty, beginning with Dr. L. Jolyon West, chairman of the Department of Psychiatry and Biobehavioral Sciences. Dr. West was then lecturing in the

East and would return within several days. Actually, I had met Dr. West at a meeting of the American Psychiatric Association some months earlier, when we both had participated on the same panel. I remembered him as a man of great vigor and philosophical depth; it pleased me that Dr. West had suggested to the dean that I be invited to join the faculty of his department as special lecturer. Dr. Mellinkoff strongly supported the invitation.

I returned to my office in New York with mounting anticipations of major changes in my life. As I mentioned earlier, almost the whole of my professional life had been spent at the *Saturday Review*. There were brief leaves of absence from the magazine for special government missions abroad. In the main, however, my life had been interwoven with the *Saturday Review* for so long that it was difficult to imagine any other identification. Though I worked in New York, I had lived most of my life in Connecticut. Was I being foolishly optimistic in believing that, at the age of sixty-two, I could start a new career at the opposite end of the continent? My wife, Ellen (officially Eleanor, but called Ellen by people close to her), was born in Utah and spent her teenage years in San Francisco. She encouraged me to believe that I would relish the challenge of the move. As it turned out, she was right—as usual.

Not long after I returned to New York from my visit to UCLA I received a letter from Dr. Mellinkoff telling me that the invitation to join the medical school as a member of the faculty of the Department of Psychiatry and Biobehavioral Sciences had been confirmed by faculty vote, and that Dr. West was looking forward to meeting me to discuss my work. That same morning I telephoned a real estate agent in New Canaan, Connecticut, and arranged for our home to be put on the market. Just saying the words "We have decided to sell" was a wrench after most of a lifetime in one place.

Ellen and I had gloried in our Connecticut house and grounds. It had ample space in which to bring up four daughters and to accommodate several thousand books and copious personal files. Ellen had as much land as she wished for growing things, the benefits of which were manifest in five or six different vegetables at almost every meal. Our daughters could entertain their friends, indoors or out, without intruding on the grown-ups.

Up until the moment I made that call to the real estate agent, there had been a quality of unreality to our decision to pull up roots; but now the threshold was crossed. A home that had been not just an abode but an organic part of our lives had suddenly become a commodity to be divested—a decision that became painfully tangible because of an infinity of packing boxes that had to be filled.

Ellen and I had heard all about the difficulty of finding homes in Southern California, to say nothing of the cost. We were lucky, however, in one respect. When we had bought our home in Connecticut more than a quarter-century earlier, it came with a fair amount of acreage. At that time land was relatively inexpensive, but its value increased sharply year by year, so that we had something to work with in finding a place not too far from the UCLA campus on the west side of Los Angeles.

Ellen and I blocked out a week for house-hunting in California. As it turned out, only a few minutes would suffice. I had known Jack Hupp, one of the leading real estate agents in the area, through Dr. Fareed, whose tennis court was both a site for some of the best tennis played outside West Coast stadiums and a social headquarters. On my visits to L.A. over the years, I had had the good fortune of playing on Omar's court, where Jack Hupp was one of the regulars. It was natural, therefore, that I would turn to Jack for our house-hunting expedition. Jack listened very carefully to our needs, put us in his car, and drove us up toward the famous Mulholland Drive that runs precariously atop a ridge that separates the "city" side from the "valley" side.

My Easterner's conception of Los Angeles had been of a vast urban sprawl, a flat city honeycombed with massively trafficked highways or "freeways," as they are formally known, a city where millions of people were climbing over one another and gasping for air. I wasn't prepared for the long open vistas, the heavily forested valleys and ridges, the large natural protected areas within the city limits, a city park recreation facility area twice the size of New York's Central Park, and residential areas hardly less countrified than New Canaan. Jack Hupp showed us a house not far from UCLA, just off the upper reaches of Coldwater Canyon and close to the Mulholland ridge. We stood on the back terrace and looked out over an unobstructed vista of perhaps twenty miles, with a series of mountains in the distance. Overhead three red-tailed hawks were executing majestic sweeps. Off to the left a sharply rising verdant hillside devoid of homes was more suggestive of the Adirondacks than of a residential section in the second largest city in the nation.

Our apprehensions about being squeezed into a densely congested and overbuilt area rapidly receded. What about the infamous Los Angeles smog? We learned that L.A. smog mainly affects the east and south regions of the city. The west side of town was vented by the opening to the Pacific Ocean. The house Jack Hupp showed us faced west and north, with prevailing winds that served as gentle coolants. Most of the rooms faced west and had large glass doors to the outside.

Ellen went from room to room, her eyes widening moment by moment. When she explored the large modern kitchen, with a center island for preparing and cooking food, and with ample cupboards and storage space, I knew we had found our new home. The price seemed most reasonable, well within the range dictated by the sale of our Connecticut home. We made an offer then and there. I doubt that more than five or six minutes had passed since we came into the house. The offer was accepted. Later, Jack Hupp told us that we had set something of a record with his company in speed of purchase.

In his *Mr. Blandings Builds His Dream House* writer Eric Hodgins cautioned buyers to be prepared for terrifying surprises, no matter how much care they might take in examining a house. Ellen and I have had many surprises since buying our home near the top of Coldwater Canyon, but they have all been pleasant. We like the solid construction of the house, the large glass doors and windows, the high ceilings, and the proportions of the rooms. We like the generous dimensions of the closets and the abundant space for pictures.

We especially like the feeling that we are living in a bird sanctuary. Hummingbirds are attracted to the plentiful hibiscus on the terrace. Robust bluejays dominate the feeders, but tiny finches practice the art of coexistence, finding the spillover from the feeder quite adequate. Doves visit us in abundance and perch passively atop the rain gutters on the roof from which they observe the passing scene. A family of horned owls makes their home high up in one of the pine trees shading the house and grounds. We welcome the daily visitations of a family of quail, seemingly oblivious to our presence as they make their journey through a lush growth of ivy, cactus, ice plants, young orange and lemon trees, banana plants, and wildflowers. Now and then we see deer darting through the trees on the uninhabited rocky hillside to the south. Coyotes are regular visitors and make their forays on Wednesday nights after we put out our garbage for regular Thursday collection. Different sections of the city have separate days for garbage pickups; the coyotes have educated themselves about these schedules and time their raids accordingly.

On my first day at UCLA, I met with Dr. West, a giant of a man physically and intellectually. He had come to UCLA from the University of Oklahoma, where he was head of the Department of Psychiatry; he set about building a great department at UCLA by attracting teachers and researchers of national stature, among them Dr. Fritz Redlich, former head of psychiatry at Yale; Dr. Milton Greenblatt, former chief of psychiatry for the Massachusetts State Department of Mental Health; Dr. Bernard Towers, whose forum invitation had brought me to Los Angeles earlier; Dr. Herbert Weiner, one of the nation's pioneers in mind-body research and author of *Psychobiology*; Drs. Lissy and Murray Jarvik, world leaders in geriatric psychiatry and psychopharmacology respectively; Dr. Daniel X. Freedman, former chairman of psychiatry at the University of Chicago and now Judson Braun Professor of Biological Psychiatry at UCLA; Dr. Ernest P. Noble, director of the Alcohol Research Center and Pike Professor of Alcohol Studies in the Neuropsychiatric Institute at UCLA, and many others.

Dr. West called his department "Psychiatry and Biobehavioral Sciences," a reflection of his conviction about the interactions of body, mind, and society. He knew that illness, physical or mental, could be the result of many things—not just what went into the human stomach but also what went into the mind; relationships with family, friends, and the outside world; ambitions, hopes, or fears. Medical science might not always be able to conquer or ameliorate all these forces, but it could at least recognize their existence and attempt to assess their general impact on any given individual. In short, Dr. West was a respecter of complexities and the need to avoid routinized diagnosis and treatment for anything as variegated as human beings. The staff of the UCLA Neuropsychiatric Institute, which he directed, represented nearly thirty disciplines, from anthropology to zoology, and its two-hundred-bed Neuropsychiatric Hospital cared for every type of neurological and psychiatric disorder in patients of all ages.

"Jolly" West was a story in himself. He had been an expert examiner in the case of Jack Ruby, the man who shot and killed the assassin of President John F. Kennedy. During the time of Ruby's imprisonment, Jolly West talked with him at length and informed the government about the mind and personality of one of the most enigmatic public characters of his time.

Dr. West was also called in by the government as a consultant in the case of Patricia Hearst, daughter of one of the kingpins of American newspaper publishing, following her arrest for a gang bank robbery in which several bank employees were killed. His interests were as broad as his experiences. He had extensive knowledge of music and could discourse on the compositions of Rachmaninoff no less expertly than on the ideas of Carl Jung. He would arrange musical evenings at his home where one could listen to nationally known string quartets or to a world-class pianist like Mona Golabek. He had a deep knowledge of the animal kingdom and nature in general and would most certainly have been at home in the company of writers like Loren Eiseley or Joseph Wood Krutch.

Like Dean Mellinkoff, Jolly briefed me about life at the medical school and the university in general. He described research projects then in progress. He introduced me to members of the faculty. He even arranged for me to play tennis, offering himself as partner or opponent—an arrangement that has been a genuine delight. Jolly is fast and agile. We play about even.

Shortly after I joined the faculty, Dean Mellinkoff created a small group from among the faculty that would serve as something of a "think tank" to meet with me for the purpose of exploring the general field of biology of the emotions. Chairman of this group was Dr. Milton Greenblatt, who, as mentioned earlier, had been chief of Psychiatry of the Department of Mental Health for the State of Massachusetts. Dr. Greenblatt had extensive experience in the field of psychosomatic medicine. I found our personal exchanges extremely rewarding.

For more than a year, and generally on a weekly basis, the members of our think tank met in a conference room in the Neuropsychiatric Institute. We were ten in number, augmented from time to time by faculty members with special expertise in the subjects scheduled for discussion.

"You're going to run into a great deal of opposition to your notions about laughter as therapy," Dr. Greenblatt told me. "Yes, I know your views have been misinterpreted and that you refer not just to laughter but to all the positive emotions. Even so, I must tell you, in all candor and friendship, that a lot of my colleagues here think you are confusing hunches and anecdotes with medical science. They support what you say about the desirability of positive emotions but see no hard scientific evidence that attitudes can make a difference. I know you attach great importance to the need of doctors to be compassionate and to take time with patients. It's important for you to know that many of these doctors don't think that there's any connection between the attitudes of their patients and their illnesses. They use microscopes; they deal with hard facts and numbers. They respect your work as writer and editor, but now you're on their turf and must play according to their rules. If you don't, they can cut you to pieces. You're the only layman on the medical faculty and you've got to recognize your vulnerability."

I told Dr. Greenblatt that my express purpose in coming to UCLA was to search for the very evidence he was calling for. I could readily understand the commitment of physicians to hard scientific research, and I told him that I would play according to their rules, as he had advised. I knew I had to objectify personal experiences and connect them to scientific evidence. But my confidence in the existence of such evidence was stronger than ever—as was my resolve to search for it. The result of Dr. Greenblatt's comments was to make me all the more determined to mount research projects that would be credible to my colleagues.

Dr. Greenblatt's friendly advice was underlined by Dr. Charles Kleeman, head of UCLA's "CHEER" program, an acronym for the Center for Health Enhancement, Education, and Research. People from all over the country would come to CHEER in order to learn how to develop a new life-style involving nutrition, exercise, and ways of coping with stress. "Chuck" Kleeman echoed Dr. Greenblatt's cautionary notes.

"Doctors are trained to think scientifically," he said. "When you talk about specific cases you know about or your own case, doctors will regard what you have said as anecdotal rather than scientific. This is something you may want to think about."

I slept uneasily that night. The friendly admonitions of Dr. Kleeman, Dr. Greenblatt, and Dr. Murphy kept echoing in my mind. Was I pursuing an obsession that went against the predominant views of the medical profession? Was I being absurd even in raising the possibility that attitudes and moods could produce actual physiological change? My own direct experience and my reasoning power, such as it was, told me that, if negative emotions could bring about biological change on the downside, the positive emotions could create effects on the upside. But the fact that such changes had not been scientifically verified could not be set aside. Yet if physicians like Dean Mellinkoff, Dr. West, and Dr. Hitzig (my personal physician) were encouraging me to proceed, perhaps I was not as presumptuous as I feared. Then I remembered that Dr. Murphy said I would not be without allies among the practicing physicians.

I also recalled Dr. Kleeman's comments about the dangers of "anecdotes." I realized that doctors and writers, generally speaking, are trained to observe life differently. Doctors are uncomfortable with anecdotes. The quickest and surest way for a doctor to discredit or disparage an account of a single experience is to label it an anecdote.

Physicians are taught to shun conclusions based on single experiences and to look for evidence based on a substantial number of cases. Consequently, their approach has to be statistical.

Writers, however, seek out anecdotes as a way of making larger statements, a way not just of capturing attention but of highlighting a point. The anecdote or individual story is the natural language of the writer. Similarly, writers tend to shun statistics. In the writer's world, statistics obscure souls. Whole lives get gobbled up by whole numbers. If nothing is real to medical researchers except as it happens to a significant number of people, nothing is real to a writer save as it happens to a single person.

The medical scientist is trained to avoid conclusions except on the basis of repeated experiences. He reasons from the general to the particular. The writer is readily attracted to individual experiences. He reasons from the particular to the general and searches through the crowd for the few faces that can become the biography of their times. In novels, the experience of a vast aggregation of human beings becomes real only as it is portrayed through the lives of a few individuals. The evils or ordeals lodged in sectors of society are best understood in terms of their impact on individual lives and have meaning precisely because readers see themselves in the story. Reality rides on direct connections from one life to another.

This is not to say that writers and scientists have nothing in common. Both groups are seekers after truth, but they search for it in different ways and in different places. For the most part, the scientist expects that, if ultimate truth is ever found, it will be in a laboratory setting, will be demonstrated, will lend itself to quantification, and will be tested and cross-tested. While not disparaging these views, the writer has a wild hunch that ultimate truth, if it shows itself, will not come tumbling out of an equation but is more likely to emerge from a lost poem discovered by a child. It will not be tested. Nor need it be.

Little by little, I adapted to my new habitat and came to learn the semantics and style of medicine. In referring to an individual case, I did so as illustrative of a principle and not necessarily as proof of a principle. In evaluating research, I looked for evidence that it applied to enough cases before accepting it as a useful new approach. At the same time, I recognized that, despite the aversion of doctors to anecdotes, they frequently draw upon individual experiences. Indeed, medical students are told by their professors that they will learn at least as much from their patients as they will from their textbooks.

I knew I could never surrender the writer's curiosity about an individual life. It was important to know not just what happened but to know as much as possible about the person it was happening to. And it was necessary for that person to feel that you were just as interested in his life and needs as in his disease.

A word about my own medical experiences might be in order here. I have had several episodes involving serious illnesses from which I had total recoveries. The first occurred when I was ten years old and was sent to a tuberculosis sanatorium. The second involved a paralyzing illness and occurred in my fiftieth year. A third occurred when I was sixty-five, two years after coming to UCLA, and involved a massive heart attack. In each case, I was aware of the importance of the will to live. At the tuberculosis sanatorium, it was possible to divide not just the youngsters but the adults into two clearly defined groups. There were the so-called realists and the optimists. You quickly found your way into one group or the other and it became an integral and important part of your life. The "realists" knew all about the ravages of tuberculosis, which, in the first quarter of the twentieth century, was still regarded as a major killer.

The young optimists at the sanatorium didn't argue with the basic facts. They knew, however, that some kids *did* come through the ordeal. So long as this evidence was real, it fed their hopes and bolstered their will to live. Nothing to me was more striking than the fact that far more optimists were able to conquer their illness than were the "realists." The difference between the two sets of patients was not merely philosophical. The "realists" tended to stay away from group activities. They seemed to me to lead a rather joyless existence. With the optimists, however, I could laugh and have fun. We shared our books. My Tom Swift series was good trading material for *The Rover Boys*. After lights were out, we read with flashlights under the army blankets. When my flashlight batteries wore out, I was able to borrow from the kid in the next bunk or down the hall. We built snowmen and had snowball fights. And we were determined to make it back all the way.

Many years later, at the age of fifty, when I was hobbled by a disease of the connective tissue, those early lessons on the importance of the will to live did not go unremembered. Added to those lessons was a new discovery that a few minutes of solid belly laughter would give me an hour or more of pain-free sleep. At that time, of course, nothing was known about the ability of the brain to produce or activate secretions called endorphins that have painkilling capabilities. Though we did not know it at the time, the laughter may have played a part in stimulating the release of endorphins. Nor was much known about the effects of moods, attitudes, and emotions on the immune system.

What seemed to me most significant in new medical research was the fast-developing concept of the human brain as a gland and not just as the seat of consciousness and the nervous system's switchboard. In addition to endorphins, with their morphinelike molecules, the brain produces several dozen secretions that have a role in asserting the body's balances and in serving the purpose of a master apothecary.

The heart attack occurred in my sixty-fifth year, and it taught me a great deal about the essential robustness of the human body and, in particular, how even a badly damaged heart can repair itself. It also taught me something about the need to avoid feelings of panic and helplessness. I realized that anxiety and depression were major enemies in overcoming or recovering from serious illness. I learned that regeneration was a basic life force and that progress could be defined as what was left over after you met an impossible problem.

Scientific research begins with a theory or even a hunch. The theory generates the research, just as research spawns additional theory. I shared my hope with Dr. Greenblatt that UCLA might become a clearinghouse for research in the glandular functions of the brain. The members of the Greenblatt group would bring copies of articles from the world medical press reporting accounts of mind-body research. These articles showed that the brain was perhaps the most prolific gland in the human body, capable not only of providing or activating its secretions but also of combining these secretions in prescribing for the human body.

I was also on the lookout for scientific evidence concerning the way human beings were able sometimes to refute the grim predictions of experts. Let me hover over this point. A dramatic case illustrating the way experts can sometimes underestimate the ability of the human spirit to rise to important challenges was the project in 1953 to bring Hiroshima survivors to the United States for surgical and psychological treatment and rehabilitation.

The U.S. State Department had opposed this project, which involved several dozen young women who had been disfigured or crippled by the atomic bombing of Hiroshima and Nagasaki. The State Department had consulted with cultural anthropologists who had confidently predicted that the project would fail because the young women would be unable to make drastic social and cultural adjustments to a strange country with its differences of language, customs, and food—to say nothing of the terrifying experience of hospitals and surgery so far away from home. The experts contended that just the fact of fear could compromise effective treatment.

It seemed to me that something was missing in the predictions of the consultants. No doubt, far-reaching adjustments would confront the "Hiroshima Maidens" on a journey to a distant country, especially in view of the medical and surgical encounters awaiting them. But they would be living with American families and responding to outstretched hands and not to impersonal situations or institutions. The American surgeons and doctors who would be treating them had come to Japan to meet them and get to know them; the surgical instruments would not be in strange hands.

Fortunately, General John E. Hull, who had succeeded General MacArthur as head of the Occupation in Japan, disagreed with the State Department experts and offered his full support for the project, even to the extent of providing his own plane for the flight of the young women to the United States.

The Hiroshima Maidens stayed two years in America. There was not a single instance of maladjustment. Not a single Maiden refused to go through with the surgery or asked to go home because of the strange food, customs, homesickness, or any of the other psychological blocks predicted by the experts. What the experts missed was the way the human mind could override statistical evidence in response to deep determination or the anticipation of a loving experience.

The experience with the Hiroshima Maidens fortified me in my convictions about the validity of psychological and emotional factors in dealing with medical problems.

The mind-body connections being explored by the Greenblatt group were running parallel to work being carried out at more than a dozen medical research centers. At the University of Rochester, Robert Ader coined the term *psychoneuroimmunology* in the early eighties to describe the interactions between the brain, the endocrine system, and the immune system. Although the term was new, the subject itself was integral with the medical tradition. Hippocrates, we've seen, regarded mind and body as part of a single organism. In 1852 two German medical researchers, Friedrich Bidder and Carl Schmidt, observed gastric changes in dogs under varying emotional conditions. And although Pavlov's name is synonymous with the conditioned reflex it is possible that his chief claim to fame should rest on his discovery that expectations could create digestive changes—a process that Walter Cannon called "psychic secretions."

Dr. Cannon was one of the key figures in American medicine in the first half of this century. Professor at the Harvard University Medical School, physiologist, endocrinologist, philosopher, Cannon had a well-balanced view of the entire field of mind-body studies. He carried forward the findings of Bidder, Schmidt, and Pavlov in his own work. One of his most interesting findings was that under circumstances of heightened emotion the spleen could increase the red blood cell population as much as 15 percent. He was critical both of pathologists who asked for hard morphological evidence showing that the mind had any effect on disease, and of mystics who overstated the role of the mind in healing. Both these groups, he said, needed an accurate understanding of the physiological processes that accompany profound emotional experiences. His substantial research identified a wide range of bodily disturbances caused by the emotions. His book *Bodily Changes in Pain, Hunger, Fear and Rage,* summarized forty-two papers by members of Harvard's physiological laboratory on the effects of emotions on the nervous system, the endocrine system, and the digestive system. Dr. Cannon was one of the most creative medical scholars and investigators of the twentieth century—and here let me digress to say that if you haven't read his *Wisdom of the Body* or *The Way of an Investigator,* you will do yourself a favor by making friends with these books.

One of Cannon's contemporaries, Fritz Mohr, the noted German medical scientist, made a pithy summary of the matter when he observed that "there is no such thing as a purely psychic illness or a purely physical one—only a living event taking place in a living organism that is itself alive only by virtue of the fact that in it psychic and somatic are united."

Among others who helped prepare the way for the psychoneuroimmunologists were medical researchers like Selye, Alexander, Meyer, Engel, Bernard, Wolf, Beecher, Menninger, Wyss, Reshauer, and Heyer, who had accumulated scientific evidence on the way the mind makes its registrations on human physiology. One of the most striking recent descriptions of this interaction is to be found in Lewis Thomas's description (in *The Medusa and the Snail*) of wart removal by hypnosis. He observed that warts "can be made to go away by something that can only be called thinking, or something like thinking." He also pointed to the mysterious involvement of the immune system in this process. These phenomena served as the basis for our special studies at UCLA.

It was clear that what our group, chaired by Dr. Greenblatt, was talking about was not attitudes or emotions versus scientific medical treatment but how to combine both in an integrated strategy. Were there forces in the human mind that could be no less powerful or useful than the physician's prescription pad? More relevant still: Was it possible that the patient's own attitude toward the illness might have a bearing on the environment of treatment?

Questions such as these came to the forefront in our discussions. Gradually, I began to perceive a change in my relationship with the physicians in the group and, in fact, with the faculty in general. Faculty members came to understand, I believe, that I was not counterposing psychological against physiological factors but was talking about the combination of the two, especially in the setting of a patient-physician partnership. As a result, my ties to the medical profession as a whole grew stronger. Increasingly, I found myself invited to give "grand rounds" at hospitals and medical schools in California and elsewhere. "Grand rounds" is the term used to describe a meeting at which entire departments of a hospital or medical school will meet with a visitor who will expound on his views and then engage in general discussion.

The benefit to me from these encounters was incalculable. I had a chance to get to know some of the leading medical figures in the United States and to profit from their views on new medical trends. I was able to learn about pioneering research efforts in the field of my primary interest—the way intangibles could work to the benefit of the patient; the way the patient's will to live and attitudes in general could be incorporated into the physician's strategy of treatment; the way confidence—confidence by the patient in the doctor, confidence by the patient in his or her own healing resources—could be seen as a vital element in the recovery plan.

The connection at UCLA quickly became an adventure in learning. I was ushered into a new arena in which knowledge seemed to be accumulating faster than it could be recorded or absorbed. The particular corner of that arena that was of prime interest to me was the way the human body responded to large challenges, especially under circumstances of serious illness. It was a process of interaction involving all the body's systems.

Within a few years after coming to UCLA, I began to feel that the quest for scientific proof that could give substance to my obsession might not be as remote as I had feared or as many medical researchers believed.

At the same time, I was aware that our quest was not without philosophical implications. We were not merely probing for new clues to internal connections but were pressing at the portals of infinite mystery. Nothing about the universe is more complex, more resistant to the penetrating powers of systematic thought, more diverse in its manifestations, more elusive in its antecedents, more electrifying in its capacities, than human life itself. The mysteries we sought to unlock were only superficially connected to arcane pathways through which ideas and emotions would move from mind to body and back again. Fundamentally, we were groping for an understanding of vital sparks, precarious fractions, the conversion of essence into process, and interactions that seemed to run the full distance from positron to planet.

We had to be careful not to allow presumptions to masquerade as knowledge. But we could also recognize the need to pursue large new propositions that might or might not contain their own verification.

2

ENCOUNTERS WITH PATIENTS

The office assigned to me was in a section of Slichter Hall occupied by the Brain Research Institute. One of my first visitors was Dr. Jolly West.

"The spring term is just about over," he said, "so your lecturing chores won't begin for several months. This ought to give you time to settle in and get acquainted with your new habitat. I suspect that your routine at the medical school will evolve. We'll give you access to research in the field of mind-body interactions. You'll probably be interested in new developments in hypnosis. You'll be getting requests to meet with the medical students individually and, unless I miss my guess, you'll be asked to meet with individual patients. This could turn out to be the part of your job that will take the most time. Doctors will probably call on you to buttress their patients emotionally. It helps patients to talk to someone who has come through serious illness."

Jolly West guessed correctly. Within a few weeks following the announcement that I had been appointed to the medical faculty, I began to receive calls from physicians at the school or in the area asking if I would try to bolster their patients' will to live, visiting with them at the hospital or their homes.

The first patient I was asked to see was a young man of thirty, a mechanic, whose joints were so badly inflamed that he could hardly walk or move his hands. I learned that he had been mistakenly arrested by the Los Angeles police on charges of drug dealing. He was of Mexican descent, and was a look-alike for a criminal who was heavily involved in smuggling cocaine across the border.

The police had a photograph but no fingerprints to go on. The young man, whom I shall call Pedro, showed me a copy of the photograph and I could understand the reason for the mistake; the resemblance was incriminating in the extreme.

Following the arrest, bail was fixed at a level completely beyond Pedro's ability to arrange. He was put in jail to await trial and became so despondent that he could only pick at his food; weight loss was rapid. A lawyer was assigned by the state to defend the young man. Shortly after the trial began, the real criminal was apprehended and Pedro was released. But the ordeal had taken its toll on both the young man and his wife. It is not uncommon for people, under circumstances of intense exasperation or frustration or rage, to become victims of their own emotions. Walter Cannon, whose name is mentioned more frequently in this book than that of any other medical researcher, provided a clear view of this process in his book *Bodily Changes in Pain, Hunger, Fear and Rage*, referred to earlier. Dr. Cannon's student and associate, Hans Selye, expanded on Cannon's findings in his own studies on stress. Selye showed how the body, in effect, could manufacture its own poisons when under siege by negative emotions, especially protracted frustration.

The young man's physician felt the medical treatment needed to be augmented by emotional fortification. Since Pedro's symptoms were similar to the ones described in my book *Anatomy of an Illness*, the physician believed that his patient might be helped by an approach comparable to my own. I met with the young man and his wife two or three times a week for about two months. He and his wife had been devotees of fried and fatty foods—an unnecessary burden on his body at a time when as many negative factors as possible had to be eliminated. My wife

Ellen helped to design a simple but nourishing diet for them with the emphasis on fresh vegetables, fruits, and fish. Meanwhile, I was encouraging them in the development of a new life-style. It was necessary to restore their faith in human beings, to become involved in good works, to be able to laugh and play. While the arrest had been unjust, they could at least be thankful that the real criminal was apprehended. After all, the authorities were not acting capriciously; Pedro did have a remarkable likeness to the photograph of the drug dealer. The mistake was discovered. Things could have been worse, much worse.

Little by little, the improvement became apparent. The young wife was the first to become depression-free. She happily took responsibility for superintending their diet. She also led the way in rearranging their life-style in order to accommodate greater exercise of the positive emotions—love, will to live, play, fun, laughter, purpose. Their physician reported consistent improvement in Pedro's physical condition, as measured by the reduction of inflammation, absence of fever, weight gain, and favorable changes in the latex fixation test (an index of the level of rheumatoid factors). Most encouraging of all, of course, was the steady reduction of pain in Pedro's joints. He was restored to his job.

Meanwhile, there were other patients I was asked to see by the dean and members of the UCLA medical faculty—patients whose despair or defeatism was impairing treatment. What I tried to do, most of all, was to enhance the environment of medical treatment. What went into the mind could create a context both for effective medical care and for the restoration of the body's own resources. To the fullest possible extent, the materials of hope, determination, and the will to live were used as vital ingredients. I knew that if patients accepted a negative outcome as being inevitable, it was likely that they would move along the path of their expectations.

I essayed the role of responsible cheerleader, passing along what I had learned from patients who *did* manage to experience remissions or to live significantly longer than was forecast. I also worked with members of the family; they were at risk because of the emotional wear and tear.

Perhaps the most striking experience of all came late in 1978 just as Ellen and I were about to leave for China. A physician in Sherman Oaks telephoned and identified himself as Dr. Avrum Bluming. He was calling about his patient, a judge, who was in the terminal stage of cancer and who was then at the Encino Hospital. He said the judge's mood was understandably bleak. All of his personal and professional life he had been known for courage, determination, and a positive outlook on life. His illness, however, had given him the psychology of fatalism. He told his wife and children that there was no hope and that he expected to die very soon.

Dr. Bluming told me that the effect of the judge's mood on the family was catastrophic. He said that the judge's seeming willingness to give up without a fight was totally out of character. The physician was worried that the judge's wife might be vulnerable to serious illness.

On the way to the airport for the flight to China, I stopped off at the Encino Hospital. Before entering the judge's room, I met with Dr. Bluming, who told me that the judge had virtually stopped eating and was resisting intravenous feeding. At the present rate, he said, it was doubtful that he would survive more than two or three days.

When I entered the room, the judge bade me sit close to the bedside. He spoke in a hoarse whisper and it was difficult to follow what he was saying, but I picked up enough to learn that he had been a long-time reader of the *Saturday Review* and had sympathetically followed its various enthusiasms and concerns.

I took his hand and thanked him and told him that few things in my life were more gratifying than to meet readers of the magazine. I asked how he felt. He closed his eyes and shook his head.

I said that Dr. Bluming had given me a briefing on his condition and that I was also concerned about his wife and sons and, in fact, about all the people who loved him.

His eyes narrowed in a way that indicated he wanted me to explain myself. I said I understood that all his life he had been a fighter for things he considered just and right.

He nodded and again he narrowed his eyes as though to find out what I was getting at.

I said that one of the things I had learned at the medical school was that the attitude of the patient had a profound effect on members of the family. Their health could be jeopardized by negative attitudes of the patient. I said I hoped he would forgive me if I said that his family was anguished by the judge's apparent defeatism. Such defeatism might seem natural in anyone else, but in the judge . . .

The judge closed his eyes momentarily. Then he looked at me and uttered just two words:

"I gotcha."

The emphasis and sense of purpose even in his whisper were unmistakable, as was the pressure of his handshake before I left.

When Ellen and I arrived in Hong Kong, the first thing I did was to telephone the Encino Hospital. Dr. Bluming went out to the nurse's station to take the call.

"Something is happening here that you'll find difficult to believe," he said. "When the nurse began to rig up the intravenous device, the judge demanded that he be given breakfast on a tray. This was done. He got the food down, and kept it down. How he did that I'll never know. When his wife arrived, he called her to the bedside, then invited her to work on problems that come up in bridge games. The judge used to be a tournament bridge player. Where he got the energy to concentrate on bridge, I have no idea.

"This isn't all," he continued. "After bridge, he asked for a robe and slippers, got out of bed, and went to the bathroom on his own. When the nurse tried to restrain him, saying she wanted him to use the bedpan, he waved her off and said he could take care of himself. He was crusty and strong-willed—just the way people had always known him."

I asked if this was any indication that the underlying situation had changed.

"Not so far as I can tell but it sure has made a difference in the lives of his wife and children. He's going to survive this weekend and then some."

After I arrived in China, we were escorted into the interior of the country where international telephone facilities were not readily available. It was not until two weeks later, when we arrived in Shanghai, that I was able to telephone the hospital again.

This time, the judge's wife went out to the nurse's station to take the call. Her voice was strong and cheerful.

"The judge's spirits have been wonderful," she said. "He has had good talks with our sons. He follows the newspapers and makes his usual witty comments. He now takes walks in the hospital corridors and chats with other patients. The ultimate outlook hasn't changed, but the general atmosphere has. We are . . . well, a lot less despondent than we were."

The judge survived for several more weeks. It was a magnificent example of how the human spirit could make a difference—not just in prolonging one's life but in bolstering the lives of others. The judge's deep sense of purpose didn't reverse the disease—the cancer had spread so widely to his vital organs that it was only a question of time before it would claim his life. But he was able to prolong his life beyond the expectations of the physician. He was also able to govern the circumstances of his passing in a way that provided spiritual nourishment to the people who loved him. He died in character. This was his gift to everyone who knew him.

Hope, faith, love, and a strong will to live offer no promise of immortality, only proof of our uniqueness as human beings and the opportunity to experience full growth even under the grimmest circumstances. The clock provides only a technical measurement of how long we live. Far more real than the ticking of time is the way we open up the minutes and invest them with meaning. Death is not the ultimate tragedy in life. The ultimate tragedy is to die without discovering the possibilities of full growth. The approach of death need not be denial of that growth.

3

LEARNING FROM MEDICAL STUDENTS

"I think you'll find that the more informal the setting, the better your rapport with medical students," Dr. Bernard Towers said. Dr. Towers had invited me to be part of his program on "Science, Law, and Human Values." He had come from England a dozen years earlier and attracted attention as a medical ethicist. He was the moderator of a medical forum at UCLA in which, as mentioned earlier, I participated several weeks before I was appointed to the faculty.

Dr. Towers helped make the arrangements for groups of medical students to come to our house for evening rap sessions. Ellen relished the opportunity to give the young people a nourishing meal instead of the hurried sandwiches or hamburgers that often comprised their evening fare.

Some of the students at these evening exchanges seemed dubious about the role of psychological factors in causing disease or in contributing to a strategy of treatment. If a bacillus was identified in a workup, they tended to believe that the way of dealing with it was clear and unambiguous: Prescribe the right antibiotic. They lived in a new world of technology designed to provide exact answers. The technology spewed out neat little numbers that all fitted together. The study of anatomy or physiology or biochemistry provided precise descriptions. Everything had a name and everything was in place.

References to the need for a patient-physician partnership, or to the communication skills of the doctor, or to medical ethics, or to the philosophy of medicine or even to the history of medicine, were regarded by some of the students as "soft," and therefore not really primary in medical education. By contrast, subjects such as physics, biochemistry, pharmacology, anatomy, etc., won the favored adjective "hard."

The reasons were not obscure. "Soft" subjects lacked precise answers. The grade-conscious student, therefore, could never be sure that the answers on examination papers would correspond to the professor's judgment. With the "hard" subjects, however, the correct number or fact was certain to lead to a predictable grade. The students therefore tended to steer away from the "soft" and gravitate to the "hard."

I was troubled by these habits of thought, for they had unfortunate lines of connection to the world beyond the medical school where uncertainties and not precise answers lay in wait at every turn and where variables characterized most of the equations. Diseases were classifiable, to be sure, but most of the patients who had them were not. In the outside world, a surprisingly large part of the factual base in medicine was vulnerable to new findings and theories. But the need of the physician to motivate or inspire patients remained constant. In the years after medical school, much of what had been regarded as "hard" turned out to be frail or faulty, and much of what had been regarded as "soft" turned out to be durable and essential.

In these evening rap sessions, I tried to express the view that medical education should produce well-rounded human beings, interested in people and not just in microorganisms, physicians who could comprehend the reality of suffering and not just its symptoms, and whose prescription pads didn't exclude the human touch.

Even as I stressed these issues, I realized that the admissions committees of many medical schools had a tendency to use yardsticks that emphasized contrasting qualities. I could understand the reasons. There were at least twenty applicants for every opening. Inevitably, this made grades the most tangible measure on which the school could base its admissions decisions. Unfortunately, the emphasis on grades had the effect of making premedical students fiercely competitive. They tended to become barracudas, chewing up each other, knowing the only way they could rise to the upper half of the class was if someone else was in the lower half of the class. The end product was not necessarily good scholarship but more often a sharpening of academic predatory skills.

Years before they filed their applications to medical schools, students would find themselves veering in the direction of numbers. They had little time to think about the world or to assess and pursue their other potentialities. They had time only to demonstrate their academic abilities, time to pursue the habits of grade-grubbing that would get them into and through medical school. This would not necessarily make them good doctors, nor would it put the emphasis on compassion, sensitivity, and respect for life as essential qualities in health care professionals.

I refer to the situation as it was in the late seventies and early eighties. Since then, medical schools have become increasingly conscious of the importance of the humanities and the liberal arts. However, too many medical students still place excessive value on numbers and grades. Emphasis on grades is apt to lead to the continued pursuit of high numbers in different forms: a better address in front of which to hang their shingles, or a specialized practice beyond the economic capacity of many of the people who most need their expertise.

If people are content to be treated by technologists, there is no reason to be concerned. But if what we want are physicians who understand the individuality of illness, who help patients to develop their own resources in combating illness, who assess all the intangibles and imponderables that affect the full functioning of a human being, then we may want to give our support to those leaders of medical education who want to broaden medical education. One such leader is Dr. Joel Elkes, professor emeritus at Johns Hopkins University, and at present professor of psychiatry and director of the Division of Behavioral Medicine at the University of Louisville, who has initiated programs to teach medical students how to maintain their humanity in the face of a mechanistic society.

It is a serious error to suppose that either admissions policy or curriculum policy are fashioned entirely outside the arena of public consensus and consent. If medical school officials are encouraged by the public to search actively for the young men and women who are capable of bringing a certain artistry to the science of medicine, then the stage may be set for wide reforms and great benefits.

These lines are written within a month or so of my tenth anniversary at the medical school. When I think of major changes that have occurred during that time, the first thing that comes to mind is the impressively large number of medical students of both sexes who intend to serve on medical missions in third-world countries. When the students had just begun to come to the house for dinner and rap sessions, most of them would speak of their intention to go into specialized branches of medicine. By the end of the decade of the eighties, the gravitational pull was to medical service in Africa, Asia, or Latin America.

An equally significant change since 1980 is the steady increase in the number of women going into medicine. In 1978, women represented about 25 percent of the enrollment. Ten years later they account for more than 40 percent, and there are indications that, within another decade, they may be in the majority.

What is true at UCLA is also true of medical schools throughout the country. The gender shift may well be the most striking new development in American medical education. The single greatest consequence of this change is an increase in the number of doctors going into general practice. For almost half a century, the trend has been to specialization, the advantages of which are represented by the availability of highly skilled physicians in dealing with intricate problems concerned with disease and disability. The disadvantages are represented by an inevitable increase in the distance between the patient and the primary physician; an increase, too, in the depersonalization that can distort the patient-physician relationship. Nor must we overlook the vast increase in medical costs that has accompanied the advent of high specialization. What a half-century ago appeared to be a salutary development in medicine—the availability of honed expertise for elusive medical problems—in time has become a series of overspecialized arenas that departed from the Hippocratic philosophy of an integrated approach to the patient. A patient who is shuttled from one medical office to another, encountering highly sophisticated technology along the way, is likely to be the recipient of a more accurate diagnosis than otherwise, but he may not be the beneficiary of an environment ideally conducive to treatment or healing.

Plainly, the trend to medical specialization has gone too far. Plainly, too, the effect of a substantial increase of women in the medical profession may well be to create a better balance between specialization and primary care. The qualities that tend to be crowded out by overspecialization—greater attention to communication needs, awareness of the importance of reassurance, increased emphasis on the need to understand the circumstances of a patient's life and not just his symptoms—I like to think that these qualities will become primary again with the increase of women in medicine.

True, I have seen no studies showing that women make more compassionate doctors than men, or that they pay more attention to communication skills than men, or that they are more adept at reassuring patients than their male colleagues, but it has been established that the best patient-physician relationships are between patients and general practitioners, whose ranks are now being swelled by women. A far lower percentage of primary care doctors is involved in malpractice suits than any other branch of medicine. The reason is probably that people are not inclined to sue someone with whom they have a close or established relationship.

In any case, the women among the medical students whom I have met at seminars or house meetings volunteered the opinion—not generally contradicted by the male students—that they may be more adept in meeting the nurturing requirements of patients than their male colleagues.

To the extent that these differences may be true, it is also likely that the related effects may involve lower medical costs. Since specialization and high technology go together, and since the combination of the two can be expensive, any trend in the opposite direction will probably bring down the costs of medical care. This is hypothetical, of course, and it will be interesting to see if such is actually the case as more and more women go into practice.

One would expect that the fact that a higher percentage of women than men intended to go into general practice would also be reflected in a higher regard by women for the so-called soft subjects—medical ethics and the medical humanities, for example. Yet the concern over grades was equally distributed over both groups.

I brooded over the tendency of the group as a whole to downgrade the very factors that were basic to effective medical practice, then hit on what I hoped would be a way of persuading them that the "soft" subjects were perhaps even more vital to a successful medical career than the subjects that lent themselves to quantification or precise answers and thus to good grades.

Most medical students intend to go into medical practice rather than research or teaching. They would therefore be competing for patients at a time when all the signs indicated a future surplus of physicians. Why not develop "hard" information about the way people choose doctors? What were the traits or style of practice that are highly valued by patients? Did patients search out the most knowledgeable doctors or did they take knowledge for granted and look for other qualities?

The residential neighborhood surrounding the UCLA campus is what the demographers call "upscale." In terms of education, occupation, and economic status, the people in this neighborhood fit the optimum profile of desirable patients. A survey of residents in this area was decided upon and a mailbox questionnaire was drawn up. In general, two major questions were asked:

1. Have you changed your physician in the past five years, or are you thinking of changing now?
2. If you have changed physicians or are thinking of changing now, what are your main reasons for doing so?

Fifteen hundred of these questionnaires were distributed and produced a 70 percent response. Obviously, the ability of the people in this neighborhood to pay medical fees would not be regarded as a disadvantage by students intending to go into private practice.

The most startling figure that turned up was that 85 percent of the respondents reported that they had either changed physicians in the past five years or were thinking of changing now. Equally startling were the reasons behind the change. When these reasons were tabulated, we discovered that people took knowledge for granted; that was what a medical diploma signified. Beyond that diploma, however, were other factors that counted heavily with patients. The questionnaire revealed that most people changed physicians not for reasons of competence but because of the doctor's style or office manner. They were troubled by insensitivity to their needs, or poor communication techniques, or by lack of respect for the patient's views, or by overemphasis on technology. Here were some representative responses:

"I left the doctor's office feeling worse than when I came, and that is not what I went there for."

"I waited almost two hours for my turn and no sooner did I get into the doctor's office than he became involved in a long personal telephone call."

"I don't think the doctor really understood what I was saying. He put me into a long series of tests even before I finished telling him my symptoms."

"The doctor smoked and I thought that anyone who had so little regard for his own health should not be entrusted with mine."

"My doctor had gained a lot of weight since our last meeting and was quite obese and I had difficulty in taking him seriously."

"When he told me what was wrong, it was so cold and matter-of-fact that I felt he didn't care whether I would get well or not."

"The doctor sent me bad news by registered mail. I suppose he wanted to protect himself but he lost a patient. I am not a legal item or oddity but a human being."

"The prescription the doctor gave me had terrible side effects he didn't warn me about."

"I don't think he respected my intelligence or my right to ask what I thought were relevant questions."

The questionnaire had an even greater impact on medical

students than I had dared hope. The information it turned up had pocketbook significance, hardly a "soft" fact in itself. It had something to do with the shine on the shingle. Whenever I presented the results of the questionnaire, whether in class or in informal supper get-togethers, I discovered that the medical students were not inattentive.

An article about the questionnaire survey appeared in the *New England Journal of Medicine* and resulted in substantial correspondence. The consensus of the letters was that while patients should not be allowed to dictate the nature of medical practice, they were the final arbiters of the quality of medical practice. And so long as patients had freedom of choice in the selection of physicians—a principle stoutly defended by the medical profession—the way they exercised their choice could be a life-or-death matter for medical practitioners.

One of my colleagues at UCLA was not at all pleased by the results of the survey, especially by the importance attached by people to compassion and communication skills. He said he was not interested in charming his patients or winning popularity contests. He said he wanted to give them the benefit of his special skills and knowledge. He said that if he himself were ill, he would select a physician who could make a correct diagnosis and who would know exactly what had to be done—rather than choose one who was adept at sweet-talk or at making friends and influencing people, or in gaining a reputation as one hell of a nice fellow.

I believe my colleague may have missed the main point. He assumed that a doctor had to be one or the other; that if he is competent he couldn't be compassionate. He seemed to believe that the science of medicine and the art of medicine functioned on different levels and in different ways.

Another colleague felt that too much emphasis was being given to communication techniques.

"What do you want me to do?" he asked. "Do you want me to hedge or dissemble in communicating a diagnosis? I have the obligation to tell what I know. My patients are entitled to the truth and I have the obligation to give it to them. My lawyer tells me that if a patient discovers that his condition was worse than I led him to believe, I could be in line for a malpractice suit that could wipe me out."

And again, I feel that my colleague may have missed the point. Certainly, he had the obligation to tell the truth but he also had the obligation to tell it in a way that did not leave the patient in a state of emotional devastation—the kind of emotional devastation that could compromise effective treatment. Medical students spend years in learning how to diagnose but only minutes in learning how best to convey it.

Question: Is it possible to inform a patient truthfully about a serious diagnosis and still leave the patient with something to hold on to in the form of sustaining hope?

Of course it is. I have been with cancer patients at the time of diagnosis. I recall one circumstance in particular. I listened to a physician as he told the truth. He put it in the form of a challenge rather than a death sentence. He was not telling less than he knew; neither was he telling more than he knew. In his medical journals he had read of hundreds of unexpected remissions. In his own practice he knew that patients who didn't deny the diagnosis but defied the verdict seemed to do better than others. And so he didn't feel under any obligation to provide any terminal date—nor would he have done so even if asked. He was wise enough to know that some people confound all the predictions, and he didn't want to do anything or say anything that would have the effect of a hex on the patient.

Unquestionably, some doctors tend to discourage discussions in which patients present their own notions about their illness. Most of the physicians I know, however, believe that the key to sound medical practice is to be found in listening carefully to the patient, however wide of the mark the theories may seem. Somewhere in the swirl of words there may be valuable clues that the physician can put to good use. Moreover, talking provides catharsis for the patient. Just in the act of expressing himself or herself, the patient's anxiety may be lessened. Since anxiety and depression are unfortunate concomitants of disease, most physicians genuinely welcome the desire and ability of patients to unburden themselves of their fears and portents.

People go to doctors out of fear and hope—fear that something may be wrong but hope that it can be set right. If these emotional needs don't figure in the physician's approach, he may be treating half the patient. The question is not now—any more than it has ever been—whether physicians should attach less importance to their scientific training than to their relationships with patients, but rather whether enough importance is being attached to everything involved in effective patient care.

The most important recent change in the practice of medicine, I believe, may be represented by the enlarged knowledge and new respect for the apothecary built into the human system. This new emphasis has been accompanied, logically enough, by increased attention to the concept of the patient-physician partnership, in which the physician brings the best that modern medical science has to offer and the patient brings an environment congenial to treatment, by which is meant confidence in the physician, a strong will to live, and a determination to get the best out of whatever is possible.

Few things are more exciting about the human body than the wide array of forces within it that are poised to do battle with invaders or abnormalities. The body possesses an immune system designed to meet the challenge of illness. Different cells in the immune system do different things. "Sentry" cells roam throughout the body, locating and identifying intruders or abnormal situations, and then summoning the body's own defenders—cells that can pry open malignant cells and inject the body's own poisons, cells that can destroy infecting agents, cells that can even arrest viruses and summon reinforcements.

It is only in recent years that extensive knowledge has been accumulating on the immune system, its component parts, and its interactions with the rest of the human body. Included in the functioning of the immune system are the brain, the spleen, the thymus gland, the bone marrow, and the lymph nodes. It is a mistake, therefore, to think of the immune system as located in any one place or as confined to any single process. It is connected to the body's vast network of organs and systems.

Similarly, if one asks, What are the primary influences on the immune system? the answer is, Practically everything. The immune system can be affected by biochemical changes in the body, by an invasion of microorganisms, by toxicity, by hormonal forces, by emotions, by behavior, by diet, or by a combination of all these factors in varying degrees. The immune system is a mirror to life, responding to its joy and anguish, its exuberance and boredom, its laughter and tears, its excitement

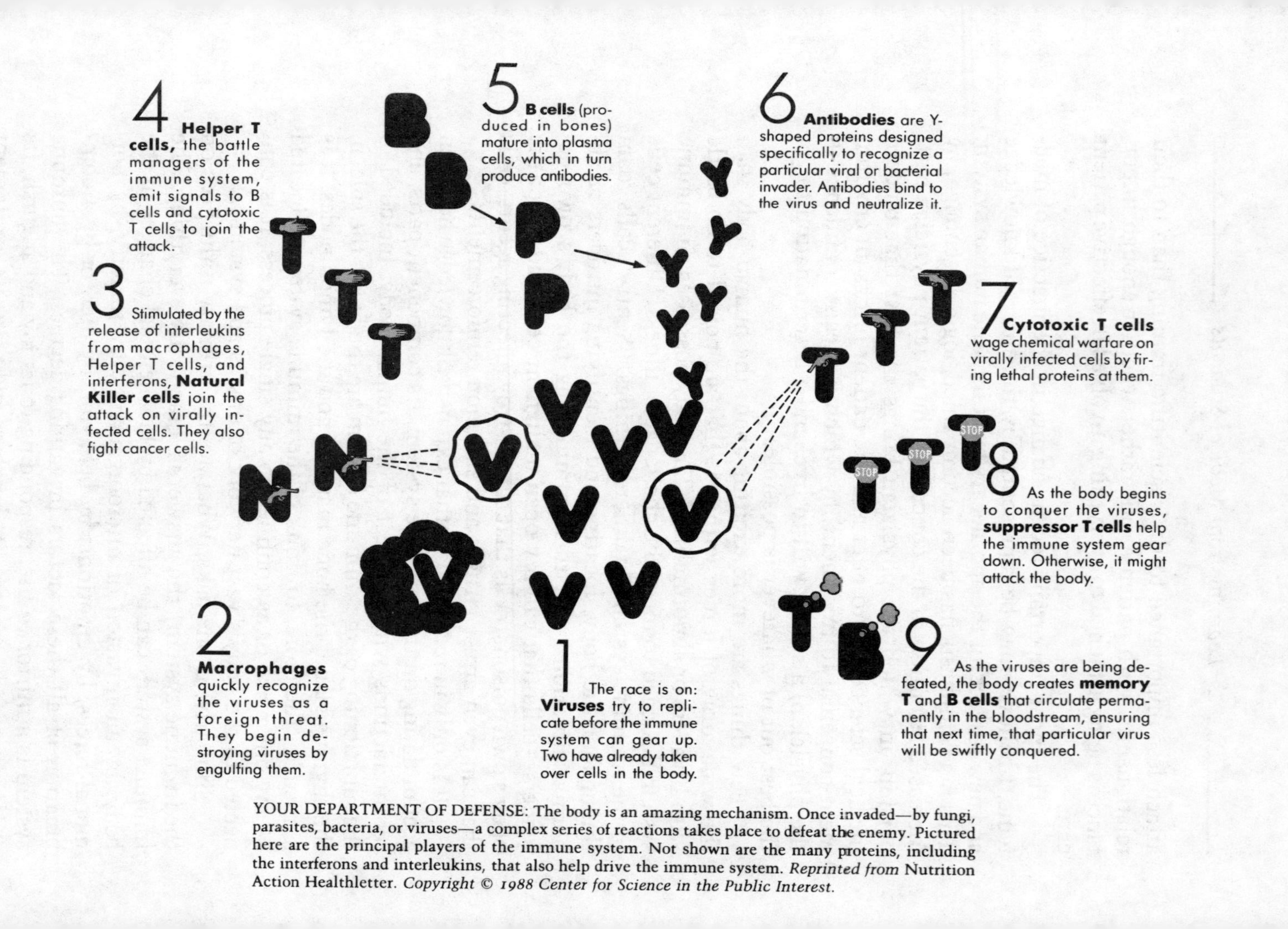

YOUR DEPARTMENT OF DEFENSE: The body is an amazing mechanism. Once invaded—by fungi, parasites, bacteria, or viruses—a complex series of reactions takes place to defeat the enemy. Pictured here are the principal players of the immune system. Not shown are the many proteins, including the interferons and interleukins, that also help drive the immune system. *Reprinted from* Nutrition Action Healthletter. *Copyright © 1988 Center for Science in the Public Interest.*

and depression, its problems and prospects. Scarcely anything that enters the mind doesn't find its way into the workings of the body. Indeed, the connection between what we think and how we feel is perhaps the most dramatic documentation of the fact that mind and body are not separate entities but part of a fully integrated system.

It is only in the past decade or two that technology has enabled immunologists to have a precise count of the different immune cells functioning under varying circumstances. World-renowned immunologist Dr. Gustav J. V. Nossal of the Walter and Eliza Hall Institute of Medical Research at Royal Melbourne Hospital in Victoria, Australia, summarizes the amazing role of the immune system: "The task confronting the natural defense system can be capsulized in six key words: encounter, recognition, activation, deployment, discrimination, and regulation." Foreign or harmful cells and substances are targeted (encountered) by immune cells that do not know in advance what they will be asked to recognize. The immune cells are then activated to respond and trigger an elaborate multifaceted defense (deployment) strategy to rid the body of the invading agents or microorganisms, all the while discriminating between the pathogen and the body's own tissues so as not to render any unwanted harm. Finally, the immune system participates in its own regulation to gauge the amount of response necessary to provide adequate protection for the body. Pioneer immunologist Dr. J. Edwin Blalock of the University of Alabama, Birmingham, appropriately describes the immune system as a sensory organ which "senses" cells and substances that are not recognized by other bodily systems.

And now the pathways along which the human mind makes its registrations on physiology are being probed more deeply than ever before. A biology of the emotions is coming into view. For example, discoveries have been made that both the neuroendocrine and immune systems can produce identical substances (peptide hormones, or neuropeptides) that influence both neuroendocrine and immune activity. The two systems also share the same array of receptors with which these substances can interact and transmit their messages. Such evidence has led Dr. Blalock to believe that the central nervous system and immune system convey similar information to each other through such hormone signals.

These facts fit in with the last article written by the late Franz Ingelfinger as editor of the *New England Journal of Medicine*, in which physicians were reminded that 85 percent of human illnesses are within the reach of the body's own healing system. Hence the importance of the expanding knowledge about the way mind and body can collaborate in meeting serious challenges.

Understandably, many physicians are apprehensive about the way this new knowledge is being misused or heralded. They perceive, and rightly so, a danger that a wide door can be opened through which all sorts of exotic and untested ideas can pass as substitutes for competent medical care. While guarding against such dangers, it is nonetheless important, as Dr. Walter Cannon said, to maintain a balanced view, one that rules out extravagant and unscientific approaches, but that at the same time fully comprehends the interactions of mind and body.

The new term, psychoneuroimmunology, has come to be descriptive of these complex interactions—interactions among the nervous system, the endocrine system, and the immune system. Now that the various other systems of human beings have been discovered and described—all the way from the circulatory system to the autonomic nervous system—the system of interactions is being charted. In that system, every action or event has an effect on the totality.

Out of the wide new research, a much clearer picture is emerging of the way ideas, emotions, experiences, and attitudes can create biological change. Some patients, in just anticipating chemotherapy, will experience severe nausea and all the other adverse reactions associated with the actual taking of the drug. Shouts of fire in a crowded auditorium can produce a wide range of physiological effects, including constriction of blood vessels and catecholamine surges capable of causing blood pressure changes and even rupturing the muscle fibers of the heart.

Our physiological reactions may be less readily apparent at some times than at others, but they are no less significant. It is well known, for example, that accountants, as the tax deadline approaches, experience elevated cholesterol levels. The same changes take place in many students as they approach their final examinations. Immunologist Ronald Glaser and psychologist Janice Kiecolt-Glaser, of Ohio State University College of Medicine, giants in the new area of psychoneuroimmunology, have found that medical students, as examinations approach, will experience reduced disease-fighting immune cells and detrimental changes in other components of the immune system.

The Glaser/Kiecolt-Glaser team has also observed immune impairment in individuals enduring chronic stress. In one study, thirty-four family caregivers of Alzheimer's disease victims were compared with thirty-four control subjects. Caregivers had lower percentages of total T cells, helper T cells, and helper/suppressor (T cell) ratios, as well as higher antibody titers (concentrations) to the Epstein-Barr virus (implying poorer T-cell control over the virus) than comparison subjects. Correspondingly, the more closely the caregiver associated with the victim, the lower was his or her percentage of natural killer (NK) cells (see diagram on page 36). However, caregivers involved in a support group, despite having been caregiving for longer periods of time, felt substantially less lonely and had significantly higher percentages of NK cells than those not involved with a support group.

In another study of chronic stress, Drs. Kiecolt-Glaser and Glaser compared thirty-eight married and thirty-eight separated/divorced women. The researchers found that women who had been separated or divorced were more depressed, and had lower percentages of NK cells, less immune stimulability, and higher antibody titers to latent viruses. A subset of women who had been more recently separated (a year or less) also showed more disadvantageous percentages of helper T cells. Even the quality of the marital relationship in women was reflected in their immune status. Comparable results were found in a study of the effects of marital separation on immunity in men.

The Glasers have also demonstrated that the reduction of stress or the enhancement of positive emotions can have the effect of boosting immunity. In studying a group of elderly individuals over a period of a month, they found that relaxation training three times a week significantly increased NK and T-cell activity, with corresponding decreases in antibodies for the herpes simplex virus and in self-reported intensity of stress symptoms. The participants receiving only social contact or no contact did not experience the same benefits.

Drs. Glaser and Kiecolt-Glaser also measured the effects of relaxation training on the immune systems of medical students undergoing examination stress. Although students in both the relaxation and nonrelaxation groups exhibited stress-related declines in some immune function, more frequent relaxation practice produced an increase in helper T cells. Unlike the relaxation group, the non-relaxation group reported significant increases in anxiety and other symptoms of distress.

Taken together, these two studies suggest that positive emotions might provide a buffer against the immunologic effects of stress—possibly reducing the risk of disease.

Obviously, psychological factors are important in physician communication. Patients who are devastated emotionally by a diagnosis are hardly ideal subjects for urgent treatment. Does this mean that the physician must conceal or alter his findings in talking to patients? Certainly not. As I emphasized earlier, what it does mean is that the physician's communication skills may be hardly less important than his other abilities in creating a stage for effective treatment. The physician who is able to connect a sense of challenge instead of a grim forecast to a serious disease and who is able to mobilize the patient's own resources as part of a total strategy of treatment may actually be getting the most out of medical science. However serious the disease, the wise physician works with the will to live and the hopes of the patient to no less an extent than he does with particularized therapies. Two thousand years ago, Seneca observed that "it is part of the cure to wish to be cured."

What is encouraging is that, little by little, medical schools are placing increasing value on the humanities. President Derek Bok of Harvard made a major educational issue out of the need for greater emphasis on the liberal arts. He has argued cogently that anyone who makes a profession out of treating people should have some idea about human nature, human culture, and human history. His voice has had a useful effect far beyond Cambridge.

Few things are more interesting about medical students than the way they change from the time they make their decision to select medicine as a career to the time they hang out their shingles as physicians.

Obviously, the fact of economic opportunity is no more absent, nor should it be, from such decision making than for other professions. But altruism bulks large nonetheless with entering medical students. They want to be of service to their fellow human beings; the Schweitzer ethic, knowingly or unknowingly, is strong.

A high sense of purpose, however, comes under increasing pressure through sheer physical ordeals, with internship the most strenuous period of all. The very qualities that are most desirable in physicians—compassion, understanding, moral commitment—are difficult to sustain under the physical hardships that are common in hospital training.

During my years UCLA, I have been privileged to visit medical schools and hospitals in other parts of the country, meeting with students and physicians at various stages in their training and their careers. The weakest link in the entire chain of physician training, it seems to me, may be the ordeal known as the internship. More specifically, I refer to the theory that it is necessary to put medical school graduates through a human meat grinder before they can qualify as full-fledged physicians. Working around the clock—and sometimes thirty-two hours at a stretch—is a common expectation of interns. The rationale seems to be that anyone who wants to go into the medical profession must be given a rigorous and systematic exposure to the realities of the physician's life.

It is reasonable to ask how the internship prepares the physician for the "realities." What if the preparation has the effect of dulling the sensitivities of the physician? What if it fosters feelings of resentment by an intern toward a patient who has a propensity for experiencing his sharpest pains at 3:00 A.M.? What kind of judgment or scientific competence can be expected of a physician who hasn't had any sleep for thirty-two hours? Studies of the effects of sleep deprivation emphasize the hazards of treating seriously ill patients under circumstances of extensive fatigue. The excessively heavy workload seems less a sampling of later challenges than an exercise in what I can describe only as disguised hazing at best and systematic desensitization at worst. Is any significance to be attached to the fact that the most articulate defense of the practice came mostly from those who, having survived the experience, seemed determined not to permit others to escape? For the most part, however, I have found that the large majority of physicians, on or off hospital staffs, see little justification for the practice and, indeed, have expressed serious reservations about it. They attribute the long working hours to the economic policies and concerns of hospitals.

Some of the most productive discussions I have had about the institution of the internship have been in the open forums accompanying the grand rounds to which I was invited by various hospitals. Not infrequently, the subject of physician-patient relationship during the internship would come up. It was recognized that the physician should make a special effort to win the patient's full confidence as well as to promote confidence by the patient in his own healing system. A compassionate attitude was described as a good way of reaching the patient.

At this point, however, the discussion would break wide open. Some physicians contended that internship didn't generate feelings of compassion toward patients. They pointed out that treating patients under conditions of pressure and fatigue is no more satisfying for the physician than it is for the patient. Against this viewpoint was the argument that having a patient with a bloodstained knife come at you in the emergency room is enough to quiet the compassionate urgings in most doctors' souls. Another physician said he was ashamed to admit that he hoped (generally at 3:00 A.M.) that his call-button-pressing patient would die before he got there. Compassion, apparently, is favored by circumstance.

Over and above these specific problems is a matter I mentioned a moment ago: To what extent do the burdens placed on the interns come more under the heading of hazing than conditioning? Is a harsh and punitive attitude by some residents toward interns an essential part of the training of young physicians? Is it possible that some residents enjoy and exploit their power over the newcomers? Does hazing of this sort reflect credit on the profession? Is it really necessary?

There is the counterargument that young doctors need to be trained to function under difficult circumstances. They must be prepared in private practice to be awakened at 3:00 A.M., in order to treat a patient in emergency condition. A great many hardships come with the medical franchise and need to be incorporated into medical education. Dr. Mellinkoff said he could understand how I would be appalled by some of the things I saw but that most physicians seemed to survive internship quite well and, indeed, felt the better for it.

I gave full weight to these facts, of course, but I was encouraged nonetheless by the changes that were beginning to emerge in internship policies. Some hospitals were leading the way in humanizing the training of interns. New York State issued new hospital regulations prohibiting continuous on-duty service by interns for more than twenty-four hours. I was especially encouraged by the efforts of the family practice and pediatrics divisions of the UCLA Medical Center to design working conditions for interns that were as much in the interests of the patients as in the interests of the young physicians.

4

THE PATIENT-PHYSICIAN RELATIONSHIP

The public announcement that I had been appointed to the medical faculty was interpreted by a segment of the public to mean that I was to serve as a medical ombudsman. I began to receive telephone calls or letters from people who had special problems or grievances. Most of these calls were concerned not with treatments or procedures but with situations that had their origins in faulty patient-physician relationships.

Example: A woman said she had undergone a series of diagnostic tests. The physician told her that he hadn't yet seen the results of the tests but he was "pretty sure it was cancer." The patient complained that the physician was being unscientific and unprofessional by offering a personal speculation of a catastrophic diagnosis. (The tests disproved the pessimistic forecast.)

Example: The wife of a cancer patient complained that the physician in charge of the case came into her husband's hospital room and spoke to them from near the doorway. There was no effort to ease the shock of what he had to say; he came flat out and said the tests were completed and that the news was "very bad." When asked just how bad, he said the diagnosis was cancer and that it was terminal. I asked the woman whether she preferred that the physician not tell the truth. She replied that she didn't object to being told the truth; what troubled her was the insensitive manner in which the truth was delivered.

Example: A woman in her early thirties had undergone major surgery for brain cancer. She was disappointed by the failure of the surgeon to telephone after she had been released from the hospital. When her headaches recurred, she telephoned the surgeon's office. He was not in; she left her name and number. The call was not returned, and she became disillusioned. "I hate to think I was treated by a cut-and-run surgeon," she said.

Example: The parents of a five-year-old girl with advanced autoimmune disease were told by the diagnosing physician that they would be well advised to lose no time in selecting the color of the satin that would line the child's coffin. Her father said he was shocked by the callousness of the remark and wondered whether the fact that the child was black had anything to do with the physician's manner. "Would he have dared to speak that way to white parents?" the father asked.

Example: A Hispanic man in his twenties brought his symptoms—occasional nausea, headaches, muscle pains—to a physician who used technical language in describing his diagnosis. One thing the patient did understand, however, was the doctor's statement that it was the kind of illness where anything might happen suddenly—blindness, stroke, paralysis. The patient felt he never had a chance to find out what was really wrong or why he was being warned. He said he was demolished by the offhand nature of the doctor's predictions—and his symptoms became intensified.

Example: One of my young colleagues on the undergraduate faculty complained that she was rebuffed whenever she sought to pursue with her doctor the diagnosis or recommendation he had given her. "He told me not to bother my head about such things, that he was the doctor and knew what to do. When I remonstrated and told him I wanted to discuss with him some things I looked up in the medical literature, he seemed insulted by my persistence and told me that if I didn't trust him I should go elsewhere. I told him that I didn't distrust him but that I wanted to have the kind of partnership with him you had with your doctor. He said if I wanted a partner I should go into business, and that he had spent ten years of his life studying how best to take care of patients, not how to be a good partner."

Example: A woman was diagnosed at one of the nation's leading medical centers as having cancer of the lung. She developed severe pains that were only partially relieved by medication, from which she suffered sundry side effects. Then it was discovered, through careful reexamination of the original X rays by other oncologists, that the diagnosis was mistaken. Following the new diagnosis, the woman's pains gradually receded and finally disappeared, even though she was gradually taken off the pills. The woman rejoiced over her liberation from the disease but was profoundly angered by having had to sustain four months of anguish and apprehension over the original diagnosis, to say nothing of the pain, which was none the less severe for being largely caused by her anxieties.

Example: The wife of a cancer patient was informed by the oncologist that her husband had from six to eight months to live. She reeled under the shock and then was directed by the oncologist to inform her husband of the terrible news. When she remonstrated with the physician, requesting that the prognosis come from him, she was told that "it is better this way." Her husband has since died—earlier, incidentally, than predicted—and she is still angry and resentful at being given the responsibility for conveying the tragic news.

Example: An eighty-two-year-old man experienced chest pains following bypass surgery. He was told that the pains were to be expected so soon after the operation and not to worry. When the pains persisted and increased, he returned to the surgeon's office and was told that the pains might be intermittent for two years. "After that," he was told, "you won't be here to worry about it."

Example: A woman complained that she waited almost two hours in the physician's waiting room for a report of a breast biopsy. Finally, she was brought into the physician's office and was informed that the biopsy was negative. As she was leaving, she was called back by the nurse and told there had been an unfortunate slipup in the records and that the biopsy had been positive, after all. She asked to speak to the physician but was told that he was behind schedule and would be unable to see her.

Example: A patient, having gone through rigorous diagnostic procedures, asked the physician what he had found. "If I were you," the physician said, according to the patient, "I would get my affairs in order as quickly as possible." Again, it was the casual and inartistic method of communication that served as the basis for the grievance.

Example: A physician telephoned the patient, saying that the results of the tests were at hand and that it was imperative for her to check into the hospital immediately. He would meet her at the admissions desk in one hour. On arrival at the hospital, the woman learned that the physician had been called away. It was more than an hour before he returned, at which time it was discovered there had been a mistake and that a room was not available. She was advised to go home and return the following morning. The patient complained that she had been scared out of her wits by the summons to the hospital and was bewildered by the casualness with which she was now told that it would be all right to come back the next day.

What most of these complaints had in common was that the patients felt they had not been treated with adequate respect or sensitivity. Next in order of magnitude were complaints about the size of the physician's or hospital's bills. Some of these patients were on Medicare or medical insurance and therefore were not directly affected by the bills; they complained nonetheless that the government or the insurance companies were being made to pay stiff fees for a multitude of tests that might or might not have been necessary.

One woman (I attach no particular significance to the fact that 90 percent of the callers were women) said she didn't understand why she should have been put through a battery of exploratory tests costing several hundred dollars for what she had been told at the outset was an inflammation of her inner ear. In her indignation, she went to another physician, who confirmed the original diagnosis in a matter of minutes. His bill was one-tenth that presented by the first physician.

A surprisingly large number of complaints I received were eventually resolved to the satisfaction of the patients. It developed that some of the patients had misunderstood the physicians or had misinterpreted the medical issues involved. I was struck with the frequency with which poor communication, rather than poor treatment, served as the basis for the complaints.

In giving these examples, it is important for me to emphasize that they are not to be regarded as a verdict on the medical profession. The very fact that I was perceived as a sort of ombudsman made it logical that only complaints would come to my attention. People who are satisfied with their physicians don't telephone me to report that fact. My experience on magazines had prepared me for such reactions. Most of the letters to the editor I received at the *Saturday Review* reflected disagreement or disapproval with specific policies or articles rather than congratulations for material in the magazine. Editors pay careful attention to critical comments, of course; but they know that a more balanced view of reader opinion is represented by the rate of subscription renewal.

People are making a serious mistake if they believe they can't find their way to caring, compassionate, competent physicians. All they have to do is to ask around among their friends. If they wanted to buy a new car, they would know exactly what to do. They wouldn't hesitate to check with friends who happen to own the make of car under consideration. There are a great many physicians who are prepared to spend time with patients, treating them with understanding and skill.

In the course of living in various places during the past half-century, I have had experience with a fairly large number of physicians; and I think of the extra measure of attention and affection they accorded my family and myself. For example, Dr. William Hitzig would drive a hundred miles during the middle of the night for a medical emergency. Most physicians no longer make house calls but some still do and are to be celebrated. I keep my own honor roll of doctors whose personal touch perpetuates a great tradition. The names on my list, in addition to Bill Hitzig, would include David Cannom, Mitchel Covel, Omar Fareed, Bob Kositchek, Ken Shine, Howard Weinberger.

In addition to the ombudsman's role, I also found myself regarded as something of a cheerleader. Increasingly, I was asked by physicians to meet with seriously ill patients whose will to live was sagging and who needed a morale boost. The physicians believed that the state of mind of a patient was not irrelevant in designing a strategy of treatment. They felt that my own experience with illness might carry some weight with patients whose depression or despondency could hardly be considered an ideal setting for treatment.

One of the questions put to Dean Mellinkoff by some physicians on the faculty during my early years at the medical school concerned the nature of my relationship with patients. Dr. Mellinkoff was emphatic in pointing out that I received no money from patients and that some patients, when told that no compensation was expected or would be accepted, would contribute funds to the medical school. Several hundred thousand dollars had been donated to the medical school in this manner. The dean could also point out that my main reward came in the form of evidence that the environment of medical treatment could be enhanced by the attitude of the patient. He took pains to correct any impression that I regarded emotional or psychological factors as substitutes for traditional scientific medical care, rather than as an integral part of a total strategy of treatment. Eventually, my situation at the medical school was generally understood, judging from the fact that the dean no longer was asked about the nature of my work with patients.

Not all the patients I encountered were referred by their physicians. Each day, there would be upward of ten telephone calls, most of them from patients who were at a low ebb emotionally and were reaching out for help. Not infrequently, the calls came from family members. Particularly memorable was a telephone call to my house from a woman in Pittsburgh who telephoned me at about 5:00 A.M. (She was operating on Eastern Standard Time and didn't take the three-hour time difference into account.)

"I need your help," she began. "My sister intends to get married. They're about to send out the wedding invitations. It's weird. What are they going to do? Send out invitations to the same list a few months later for funeral services? My sister has cancer. There's not much hope. She's not being realistic. Maybe if you speak to her it might make a difference. I know she's been a subscriber to the *Saturday Review* and has read some of your books. You can really help us."

"Does your sister know she has advanced cancer?"

"Yes."

"Does her fiancé know she has advanced cancer?"

"Yes."

"And they still want to get married?"

"Yes."

I said I would be glad to telephone her sister, which I did. But I didn't deliver the requested message. Instead, I congratulated the sister and told her how inspired I was by her decision to get on with her life and to reach out for the best. I said I had the utmost admiration for her and her fiancé and wished them both well. The couple proceeded with their wedding plans.

The wedding took place as scheduled. The woman went into remission. Once or twice a year I would telephone the woman to inquire about her health. After ten years she is still in complete remission. Her physician says it is difficult to convince him that her decision to move ahead resolutely didn't have something to do with the retreat of the cancer.

Not all the early-morning telephone calls turned out so well. One such call came from a woman in Massachusetts who said she had gone abroad. Now, on her return, she discovered that her son was scheduled for surgery. She felt the operation was unnecessary.

"What is the surgery for?" I asked.

"Testicular cancer."

"Why do you believe the surgery isn't necessary?"

"It could not have come about that suddenly. He's my son. I would have known that something was wrong."

"Have you spoken to the physician? Surely you are in a position to postpone the operation if it's unnecessary."

"It wouldn't work. My son wants to go ahead with the surgery."

"How old is your son?"

"Thirty-three."

"Don't you think your son is old enough to make that decision?"

"Yes, and he would agree with me if it weren't for his wife. She dominates him."

"Have you spoken to his wife?"

"We're not on speaking terms."

I told the woman I hesitated to offer advice but a specific medical decision was involved here and it might be unwise for her to intervene in a situation as serious as testicular cancer. At least she owed it to herself to find out from the physician exactly what was involved in this particular case. Surely she didn't want to deprive her son of what might well be lifesaving treatment.

"Is testicular cancer that serious?" she asked. "How do I know he really has testicular cancer?"

"That's why it might be a good idea to talk to your son's doctor. If you want to, you can get your own doctor to call the surgeon. He can ask for the diagnostic reports, then evaluate them for you. Perhaps you had better ask your son if he is willing to have you do this."

"Good idea. I'll call him at his office when his wife isn't around."

There is nothing if not variety in the kind of cases I am asked to see. A well-known motion-picture actor telephoned to say he had pains in his chest. He said he had gone to a cardiologist at the recommendation of his physician. The cardiologist examined him carefully, put him through an exercise test, then said there was a strong probability that he would need bypass surgery. First, he would need an angiogram to determine the extent and location of the arterial blockage. The actor said he was terrified by the entire business and found himself thinking and acting like an invalid.

For obvious reasons I can't disclose the name of the actor. But his appearance on the screen is so athletic that it was difficult for me to imagine him acting like an invalid.

When we met the next day, I could see the evidence of the man's anxiety. He was pale; he spoke with a slight tremble; he sat forward in the chair and his hands worked nervously on a pencil. He said he had always prided himself on being in superb physical shape. In fact he had to be, he said, because this was what his work required. But now he felt that the world was coming down around his ears. He hated the idea of open-heart surgery. Apart from the terror of the operation, he thought there was no way to keep the news from getting out.

"When that happens," he said morosely, "it's curtains. There's no way you can play a romantic lead if moviegoers know of your limitations because of a bad heart. Man, I'm worried."

I asked if he had any previous heart symptoms, or whether there was a family history of cardiac problems.

"That's just it," he said. "I've never had any symptoms before. No one in my family has had the slightest heart problem. What happened is that two weeks ago I began to feel something strange in my chest. When I took my pulse I discovered I was missing a beat now and then. I called my doctor. He told me to come right over. He took some tests, told me not to worry, but wanted me to see a cardiologist who would give me more specialized tests. The cardiologist confirmed the fact of an irregular heartbeat. He scared the stuffings out of me. He said that my irregular heartbeat was the result of deep underlying cardiac problems. He said my treadmill test showed that I had cardiac insufficiency and blocked arteries and that I would need an angiogram. That way he could find out all about the blockage and could draw up a plan for the bypass surgery. That's why I'm so scared. I don't think I could survive it. I wish I knew what to do."

I told the actor that he might want to get a second opinion. Dr. Kenneth Shine, who had been my primary cardiologist at the time of my heart attack in 1980, had been made chairman of the Department of Medicine at UCLA and his private practice had necessarily been curtailed. I told the actor about Dr. David Cannom, who had been most helpful to me personally and who had had considerable success in raising the confidence level of patients as an integral part of treatment. Then I asked the actor to describe an average day in his life. He did so. What I found especially striking was that each afternoon he would spend at least an hour in the sauna. I asked about the reason for the sauna.

"This way, I don't have to worry about my weight. I like to eat. I like candy. I discovered I could eat what I wished because the sauna kept my poundage down."

Forced weight-reduction in a sauna, especially if repeated every day, can be hazardous. Profuse perspiration can result not just in dehydration but in loss of minerals. For example, the body has to maintain an "exchange" of sodium and potassium. If the loss of these minerals is severe, the result can affect the electrolytes in the body, one manifestation of which can be a disruption in the electrical activity of the heart. An abnormal heartbeat, or arrhythmia, can be one of the effects.

I asked the actor if he had reported the fact of his daily saunas to the cardiologist.

"No," he said. "We didn't talk very much. He put me right into the tests."

The actor accepted my suggestion to obtain a second opinion from Dr. Cannom about the proposed bypass. After a lengthy discussion and various tests, Dr. Cannom advised against the bypass surgery. He also advised the actor to be more moderate about the saunas and to take adequate mineral supplementation—not just sodium and potassium but magnesium, calcium, and various others. He told the actor that proper nutrition was essential and advised against foods heavy in fats, sugar, salt, and nitrates.

Since the tests confirmed the fact of arterial blockage, Dr. Cannom suggested a new technique, angioplasty, far less drastic than bypass surgery. A small balloon would be threaded into the faulty arteries, then inflated. In this way, the channels would be widened and could furnish the heart with the oxygen it required.

When I saw the actor after his visit with Dr. Cannom, he was a changed man. The tremor in his voice had disappeared; his confidence was restored. He no longer had the feeling that at almost any moment he might topple over. He knew that his career was no longer in serious jeopardy.

A somewhat similar case came to my attention a little more than a year later, this one also involving a well-known actor. The actor, whom I shall call Robert, had had no symptoms. He had gone in for a regular checkup only to discover that the tests revealed he had had a silent heart attack. The cardiologist to whom he was referred confirmed the diagnosis. The effect on the actor, who had always been fearful about his health even under normal circumstances, was pure panic. He reported for work but was so tremulous that he had to leave the motion-picture set.

When I met with the actor, he said he feared the worst. His family doctor had told him that the treadmill test indicated the need for additional tests and that an angiogram was necessary. This advice escalated the panic into something approaching an anxiety neurosis, for the actor was not unaware of the connection between an angiogram and bypass surgery. He knew that not every angiogram procedure resulted in surgery but that the large majority of people who went in for angiograms were told to stay for the surgery.

As in the previous case, the man was in a condition of total despair. I tried to reassure him by telling him of the other actor, who was now fully functional. I also suggested that he consult Dr. Cannom before reaching any negative conclusions about the condition of his heart. It was important for him to know that the heart was a remarkably durable organ and that, contrary to the general impression, a damaged heart could be reconditioned. Dr. Cannom had had a great deal of experience with such cases.

A few days later David Cannom telephoned.

"I've examined your friend," he said. "I believe his emotional responses to the previous tests may have compromised the results. Just putting him through the thalium scan produced a vagal disturbance. This doesn't mean he doesn't have a real underlying condition. At least two of his arteries are badly blocked. But his fears during the test made the results worse than they actually were. Ordinarily we'd suggest angioplasty but I'm afraid that his anxieties during the procedure could touch off an episode. I told him about patients like yourself who suffered heart muscle damage or blocked arteries, or both, but who are now living fully active lives. I said I didn't know whether he was up to the kind of discipline required to repair his heart but I would be glad to draw up a battle plan if he was willing. He'll have to change his entire life-style. He will need daily graduated exercise. He will have to go on a rigorous new diet."

The story has more than a merely satisfactory ending. Robert applied himself conscientiously to the new regime. He held to a diet low in fats and low in salt. He stayed away from refined sugar and refined flour. He walked five miles a day. He changed his life-style to allow for creative relaxation.

At the end of six months, Dr. Cannom telephoned to report on Robert.

"We've just put him through the same tests that gave us such worry the first time," he said. "The arteries that were blocked then are still pretty well blocked now. But here's the good news. The heart has made its own bypass around the blocked arteries. It's a beautiful bypass and the heart is getting all the oxygen it needs. He's in fine shape and his heart ought to carry him a long way."

Robert looked and acted like a man completely remade—physically, emotionally, spiritually.

Several lessons are to be learned from these two episodes. The first is that full communication between patient and physician is indispensable not just in arriving at an accurate diagnosis but in devising an effective strategy for treatment. The second is that a damaged heart is not forever and can repair itself. The third is that fear can exaggerate a negative diagnosis. The fourth is that determination and discipline in creating and following a new life-style, together with expert and compassionate medical care, can actually help to create nature's own bypass.

Let me dwell for a moment on the first of these lessons. Consider the actor whose heart function was impaired because of his daily sauna and who did not inform the examining cardiologist because "we didn't talk very much," and who was put right into the tests. This instance is an illustration of the way technology can cut into the time for essential talking and listening. Little wonder that important clues are sometimes overlooked.

Doctors are not entirely to blame for spending so little time directly with patients. The central fact here is that doctors don't get paid for talking to patients. In a medical economy dominated by third-party paymasters—insurance companies, the government, health plans, etc.—the harsh reality is that doctors get paid mostly for tests and procedures. It is not surprising, therefore, that patients should be subjected to a multitude of encounters with expensive medical technology, not all of which is essential or without risk. Finally, it helps to explain why medical costs have skyrocketed in recent years. Whether in physician's offices or in hospitals, tests now account for the major portion of the physician's fees.

Even more serious, of course, is the reduced time for the careful questioning by the physician that has held such a high place in medical tradition.

There comes to mind an Englishwoman who had consulted a neurologist in Los Angeles. The doctor could find no organic cause for her symptoms—occasional fainting, dizziness, tremors in head and hands, depression. He hoped I might do something to shore up her sagging spirits.

I met with the woman and learned that she had been suffering progressively from her assorted symptoms for at least five years. She had been to see several neurologists in London, none of whom could find a cause for the symptoms. She decided to come to Los Angeles because of its reputation as an outstanding diagnostic center. Her tremors mimicked Parkinson's disease symptoms but the tests failed to confirm that diagnosis. She said she became depressed over the failure to locate the reason for her symptoms. She was profoundly disturbed by her inability to maintain her responsibilities as wife and mother. She had had to relinquish her position as a section supervisor in a London department store.

I asked the routine questions about her daily life. I asked about medication. She said she had been off drugs for at least three years.

"Are you on painkillers?"

"No."

"What about relief from depression? Do you take anything at all?"

"Oh, yes," she said, "I take Valium."

"How much do you take?"

"Usually, three or four pills a day. If my depression is very heavy, I sometimes take five."

"What strength?"

"Full strength."

"How long have you been taking Valium tablets?"

"Maybe seven or eight years."

"Why did you take it in the first place?"

"I had been feeling depressed about a family situation and my doctor gave me a prescription for Valium."

"Did you get a separate prescription each time you wanted a refill?"

"Oh, no," she said. "The pharmacist would telephone the doctor and he would authorize a refill over the phone."

I asked the woman whether she reported the fact that she was taking Valium regularly to the English neurologists she had consulted for her tremors and other symptoms.

"No."

"Why not?"

"They didn't ask me."

"Didn't the American neurologists ask you if you were taking any medication?"

"Yes."

"And did you tell them you were taking Valium?"

"No."

"Why not?"

"Valium isn't really medication. Isn't it something like vitamins?"

"No, it's medication. Surely you must have known this when your doctor gave you a prescription for it."

"I guess it was so easy to renew the prescription that they didn't really seem like medication."

I knew the woman could tell from my questions that I thought there might be some connection between the tranquilizer and her symptoms, so I suggested she discuss the matter with her doctors.

Valium is a powerful drug and can produce tremors and slurred speech if taken in large doses over a prolonged period. It can also be addictive. It can even deepen the depression it is supposed to allay.

I telephoned my friend in neurology and reported my conversation with the English lady.

"Good Lord," he said. "She telephoned only yesterday to say that she had decided to fly back to London immediately."

"Did she say anything about Valium?"

"No, and it worries me. She's probably addicted and if she tries to go off it cold turkey she can have serious problems."

This is exactly what happened. A telephone call to London from my friend produced the information that the woman had gone to the library in Los Angeles, obtained a copy of the PDR—the *Physicians' Desk Reference*—looked up Valium, read the material under "Contraindications," and saw a specific reference to the fact that the drug could produce symptoms similar to those experienced. She exulted in this discovery, feeling that the discontinuation of this drug would be the answer she was looking for, and flew home to share the good news with her family. She had gone off the drug before she left Los Angeles. By the time she arrived in London she was experiencing severe chills and other acute withdrawal symptoms. If she had been a heroin addict and had stopped abruptly the effects could hardly have been less drastic. Fortunately, she telephoned her English doctor, who pieced the information together and drew up a plan for gradual reduction and cessation of the drug.

Two months later, the English doctor wrote to say that his patient's Parkinson-type symptoms had disappeared.

I learned several things from that episode. I learned why an increasingly large number of physicians are wary of authorizing prescription refills over the telephone. I also recognized more forcibly than ever how important it is for the examining physician to push and prod and, if necessary, to dredge out of a patient the basic facts that are essential to an accurate diagnosis and a plan for treatment. The defense—that doctors don't have enough time—points to one of the most important problems in American medicine today. Since doctors don't get paid to talk with patients, technology is the great provider. However advanced, technology is no substitute for direct exchange between physician and patient. Technology can supply measurements; only the patient can supply essential background.

For the patient, it may be useful to write out in advance of seeing a new physician exactly what medications have been taken or are being taken and in what quantities. In the pressure of a visit to the doctor's office, it is easy for the patient to overlook information essential to proper treatment. A list of things can help focus the physician's attention on basic factors.

The experience with the London patient was a useful point of reference two years later when a man in a wheelchair was brought to my office by his wife. They had come in response to the suggestion by the man's physician that he could use a boost to his spirits. Severe tremors affected head, hands, and feet. He was drooling. His speech was so badly slurred that, concentrate and strain though I might, I was able to understand only every third or fourth word. His wife handed me the medical record. The diagnosis referred to Parkinson's disease.

The examination leading to the diagnosis had taken place almost a year earlier. The prime medication that had been prescribed was L-dopa. In response to my question, she said he was taking no other medication.

Mindful of the reply of the British lady to the same question, I asked about Valium specifically.

"Oh, yes," she replied, "my husband's been on Valium for several years at least."

When I commented that Valium was a medication, she said

she thought it was only a way of making her husband feel better.

"Did his present physician prescribe the Valium?"

"No, that was done by another doctor."

"Did you tell your present doctor about the Valium?"

"I'm not sure. I don't think so. We just assumed they knew it was something my husband had to take."

The husband nodded in agreement.

I suggested they should inform their physician about the persistent use of the Valium over several years. We also discussed the need for a good quality of life; for a full program of creative activities that would take his mind off his health problems and that would offset his depression; for a high level of nutrition; for a blazing determination to meet the challenge of the illness; and for strong confidence in the physician and in his own resources.

Three months later the couple returned. The man was still in a wheelchair. But the difference in his condition was spectacular. The tremors in head, hands, and feet had disappeared. Speech was still slightly impaired but I had no difficulty in grasping the meaning of what was being said.

The patient said that his doctor had taken him off Valium gradually. He said he had begun to notice the difference after only three weeks. His drooling decreased. He was able to spend less time in his wheelchair. He began to do walking exercises in the swimming pool.

"Look at this," he exclaimed and he stood up by himself. Then he turned around and took four or five tentative steps.

"My doctor says I'll be out of the wheelchair altogether in about two months and that I'll be completely well again. It's wonderful."

The episode caused me to reflect on the widespread use of Valium and other tranquilizers. Valium is the most frequently prescribed drug in the United States. More than 70 million prescriptions are written for it annually. Obviously, there are times when tranquilizers fill an important health need. The circumstances in a person's life may be oppressive and not subject to ready correction. A tranquilizer can sometimes represent the difference between suicide and survival. Instead of being regarded as a stopgap, however, while other approaches of a psychological or psychiatric nature are employed, the drug not infrequently is routinized. After a while, more and more of the drug is required and has decreasing effect, especially if the underlying problem continues. Meanwhile, the patient becomes addicted to the drug.

Unmonitored tranquilizers can be damaging to the nervous system, as in the case of the British lady and the man in the wheelchair, and they can impair the immune system, weakening the body's defense against the wide range of illnesses that lie in wait for an opportunity to take hold. According to Dr. Henri Manasse, dean of the College of Pharmacy at the University of Illinois, within every hospital there are patients who have been admitted due to drug misuse (including intentional substance abuse) and there are likely to be patients who are on an extended hospital stay or who die as a consequence of drug therapy. Every medication carries with it a risk—some drugs are dangerous enough to cause major physiological problems. Dr. Manasse reports that although individual hospital studies have suggested that 10 percent of hospital admissions are attributable to improper drug use, regional reporting systems are needed to validate such studies.

Antibiotics may be the greatest single advance in medicine of the twentieth century but their widespread use is not without problems. With the passing of years, the attacking microorganisms have become increasingly resistant to existing antibiotics, leading the pharmaceutical companies to develop ever more potent drugs. The war between the microbes and the antibiotics spirals upward and upward. But the body's ability to cope with powerful chemicals is not infinite. The more powerful the antibiotics become, the more inimical they are to the body's natural bacterial environment, hormonal functions, and even to the immune system. The physician always has to weigh hoped-for gains against possible dangers. Nothing is more important than to keep track of patients after they leave the physician's office, especially if the drug being used has penalties attached to it or if the patient seems to be developing drug dependence.

Cne thing is clear: The use of antibiotics for casual illnesses is no longer in favor with many physicians.

I discussed the general problem of medication with Dr. Kenneth Shine, who in 1986 succeeded Dr. Mellinkoff as dean of the UCLA School of Medicine when Dr. Mellinkoff retired from the deanship. He said it was the same old story. Physicians were very busy and weren't always able to keep track of all patients on medication. He said that constant monitoring was particularly important in the case of patients who had deep emotional problems and were on tranquilizers. Many of their tranquilizers were addictive and became a regular part of the daily routine—with disastrous effects in some cases.

Dean Shine said that many patients on tranquilizers or antidepressants and other drugs have learned the trick of asking their pharmacist to telephone the physician's office for authorization to renew the prescription. He also pointed out that many nonprescription drugs present similar problems. Sleeping pills, cough medicines, even aspirin tablets can carry addictive and other hazards.

It is difficult for many people to recognize that aspirin or drugs like it are powerful medications and should be taken only on the physician's instructions. The general notion exists, fostered largely by advertising, that aspirin is not only harmless but highly beneficial on a daily basis. Aspirin is a valuable drug for cardiac patients who need to be protected against blood clots. It is an effective blood thinner. But those same qualities can cause internal bleeding. Aspirin can inhibit platelet formation and have a deleterious effect on collagen. It is irresponsible, therefore, to say that *anyone* can take aspirin regularly.

The only reason aspirin is not a prescription drug is that it came on the market long before the present standards were adopted. There is little doubt that, if aspirin were a new drug and were submitted today for FDA approval, it would not pass the test for prescription-free availability.

Sometimes, people will move to a different area, develop symptoms, see a new physician, and neglect to report that they are taking drugs. Not infrequently, they will be given prescriptions that sometimes are antagonistic to the old drugs. One patient I encountered left for Europe shortly after being given a prescription for reducing his blood pressure. He traveled from place to place in Europe, becoming progressively afflicted with a wide variety of symptoms. He was treated by at least half a dozen doctors in as many places. By the time he returned to California he was on seven different drugs, had lost about thirty pounds, and was experiencing severe kidney problems, apart from pains in the chest, abdomen, and legs. His regular physician quickly identified the cause and carefully took him off all drugs. It was three months before the man was able to resume his normal activities.

Drugs to bring down blood pressure or to deal with certain kinds of cardiac abnormalities have to be carefully supervised. A physician I know in Los Angeles was almost killed in an automobile accident. He was on a drug to lower his blood pressure and he blacked out while driving. He had mistakenly routinized the drug without also routinizing the taking of his blood pressure. The antihypertension drugs he was taking, if not closely watched, can have a wide number of deleterious effects, including impaired kidney function and muscle weakness.

It is not true that the *only* way to deal with high blood pressure is through powerful medication. Overweight and high blood pressure are ready companions. Attention to one's lifestyle, especially in the direction of reducing emotional tensions, a modest but regular program of daily exercise, a diet low in salt and sugar and reasonably free of fatty meats and fried foods, and plenty of good drinking water—all these are useful and indeed essential. These can help bring blood pressure within acceptable limits. But the quick-fix tendency in American life is so strong that patients look to doctors for the convenient pill that can do the entire job without requiring the reorganization of one's life.

One patient I visited complained that his head felt as though the main wires had been crossed or pulled together. He said he was suffering from a feeling of disconnections, and the harder he tried to get the world around him in focus, the more jittery he became.

When I asked about his medication, I was handed a long list totaling fifty-six pills or capsules a day. Included in the inventory were diuretics—pills to reduce water in the body—but there was nothing to compensate for the loss of minerals that accompanied water loss. Potassium is especially important because of its role in maintaining the body's electrical system. Magnesium has a key role in regulating heart rhythm.

What was most striking of all was the heavy inclusion of tranquilizers, which alone could account for the patient's extreme jitters and feelings of disorientation. (It developed he had been on tranquilizers for eight years.)

I brought the list of medications to the dean's office for evaluation. Dr. Mitchel Covel, associate dean, reported that the rationale for the drugs connected to the patient's heart condition was clear enough but the assortment of other pills could turn anyone into a zombie. In fairness to the primary physician on the case, it should be made clear that the patient himself continued to include the tranquilizers and painkillers in his daily regimen. In any case, the hazards of copious medications were dramatically apparent.

We live at a time when progress in overcoming illness is equated with pills. If something is wrong, we seldom get at the underlying cause, especially if it means strenuous effort. We regard the doctor as the miracle man who can wave his prescription pad over us like a magic wand and provide a presto remedy. This is not the noblest exercise of the physician's skills. Unfortunately, the public's short-cut notions of how to get rid of illness or pain not infrequently put the physician in a position of meeting the demand for prescriptions or seeing his patients go elsewhere.

These observations are not original. In his lectures to medical students, Dean Mellinkoff would give the greatest emphasis to the need to avoid the quick fix, for disciplined and creative attention to what the patient was saying, and for using their own judgment at least as much as technology in arriving at a diagnosis. I saw a good example of this one day when the dean invited me to join him on medical rounds.

We met with residents and interns for the purpose of discussing interesting and instructive cases that had come to the UCLA hospital. The specific case at hand concerned a Guatemalan male, age thirty-four, diagnosed as suffering from a combination of tuberculosis and inflammation of the liver. The patient had unremitting fever and was not responding to treatment. Additional diagnostic tests were scheduled, but not much hope was held for his survival.

Dean Mellinkoff inquired about the patient's diet and learned that the nutritional level was very low because the man had no desire to eat. Consequently, the patient was receiving intravenous supplementation.

After asking several more questions about the patient's medical history and general background, Dr. Mellinkoff asked to see the patient. When he came to the patient's room with the staff members, it was learned that he had been taken in a wheelchair downstairs for X rays.

"I just hope they gave him a specific time appointment for the X rays," the dean said. "He's in no condition to have to wait on line, even if he is in a wheelchair. Let's go downstairs and see how he is."

As the dean feared, the Guatemalan, wan and thin, was huddled under a blanket in his wheelchair in the hallway, outside the X-ray room. If the dean was displeased by the reality of his foreboding that the patient would have to wait, he gave no sign of it. He approached the patient and, in Spanish, introduced himself and, very gently, asked how he felt.

The patient, obviously flattered by the presence of the dean, smiled and said everything was all right. Then the dean asked if he was able to eat his meals.

The patient mumbled something to the effect that he was not hungry.

"Are you getting food you like?"

The patient said nothing but his expression indicated that he found it awkward to answer the question.

"Do you get the kind of food you have at home?"

An ever-so-slight shake of the head indicated the answer was negative.

The dean put his hand on the man's shoulder and his voice was very soft.

"If you had food that you liked, would you eat it?"

"Sí, sí," he said.

The change in the appearance and mood of the patient couldn't have been more striking. The exchange with the dean was reassuring and therapeutic.

The dean looked over at the interns and residents. Nothing was said but it was easy to tell that a message was being sent and was also being received.

Back in the conference room the dean asked why the Guatemalan wasn't getting food he could eat.

One of the residents ventured that "we all know how tough it is to get the kitchen to make exceptions."

"Suppose," the dean countered, "you felt a certain medication was absolutely necessary but that our pharmacy didn't carry it, would you accept defeat or would you come down as hard as possible to insist the pharmacy meet your request?"

"Probably the latter," the resident said.

"Very well," the dean replied. "You might want to try the same approach to the kitchen. It won't be easy but if you want more troops on your side, I'm available. Meanwhile, let's get some nourishment inside this chap as fast as possible, and stay with it. Incidentally, there must be someone among you who can speak Spanish. If we want to make real progress, we're going to need effective communication."

Three weeks later, Dean Mellinkoff informed me that the Guatemalan had left the hospital under his own power. He had responded beautifully to treatment. He was gaining weight. The disease in both lungs and liver had receded, the fever had disappeared, and the prospects were good.

I rejoiced in the news but also recognized that there might have been an entirely different result if the dean had not intervened. The lesson to be learned was that there was much more to effective treatment than medication, and that good communication and reassurance were vital parts of any strategy for recovery.

Nothing I have learned in the past decade at the medical school seems to me more striking than the need of patients for reassurance. Thinking back over the many hundreds of patients I have seen in hospitals or who have come to my office—some of them from far-off places like India, Pakistan, Greece, France, and others from across the United States—I am forced to recognize a vital connection between reassurance and effective medical care. Illness is a terrifying experience. Something is happening that people don't how how to deal with. They are reaching out not just for medical help but for ways of thinking about catastrophic illness. They are reaching out for hope.

People tell me not to offer hope unless I know hope to be real, but I don't have the power not to respond to an outstretched hand. I don't know enough to say that hope can't be real. I'm not sure anyone knows enough to deny hope. I have seen too many cases these past ten years when death predictions were delivered from high professional station only to be gloriously refuted by patients for reasons having less to do with tangible biology than with the human spirit, admittedly a vague term but one that may well be the greatest force of all within the human arsenal. The human spirit can't be diagrammed or dissected; it can't be seen by tomographic scanners and it can't be represented by numbers on a medical chart. Yet it is the single most identifiable feature of human uniqueness. Unless it is understood and respected, all other facts are secondary.

Reassurance is not a Pollyanna concoction aimed at deception. It is not a verbal tranquilizer for creating a mood of synthetic calm. It is a way of putting the human spirit to work; a way of respecting the desire of a patient to confront a new challenge; a way of summoning all one's strength and resources in the most important fight of one's life. No one would argue against the logic and necessity for a patient to reach out for the best medical help available. Why, then, argue against reaching within oneself for the best that the human apothecary has to offer? Reassurance and hope are ways of activating that apothecary.

The wise physician doesn't minimize the seriousness of the illness; he presents it as a challenge that calls for the best both doctor and patient have to offer. Instead of dwelling on all the melancholy possibilities, he offers a plan of battle in which the patient has an active role. The physician stays in close touch with the patient, regarding morale as an integral part of sound treatment. The physician also provides the patient with a comprehensive list of things to do in addition to medical compliance. The list includes major attention to nutrition; exercise to whatever extent is possible and desirable; freedom from unusual stress, or coping with unavoidable stress at home or at the workplace; the need for a good quality of life; ways of combating depression and anxiety; and suggestions for participating in groups of patients confronting similar problems, enabling the patient to come into possession of powers that he never dreamed were within reach.

No matter how small the chance of recovery may be, reassurance is a lifeline that connects the physician and patient in a joint venture. The wise physician knows that reassurance can speed up recovery. He knows that the uncertainty and anxiety that accompany even minor ailments can magnify pain and slow the healing process. Illness gives substance to apprehension, converting fear into panic, with all the disease-intensifying factors that panic can produce.

Especially poignant for me was a telephone call from a physician whose three-year-old daughter had just undergone surgery for a brain tumor and was now under intensive care at the Los Angeles Children's Hospital. The physician said that he and his wife were desolate because the child was unable to speak or to have freedom of movement of arms, hands, and legs. He frankly said he didn't know how to cope, and he didn't even know how to console his wife. He was groping for reasons to be hopeful. His experience as a physician told him that, even if there should be no recurrence of the malignancy—it was an astrocytoma—the child might never speak again or be able to move about.

I have no competence in working with seriously ill children, but he was reaching out for the kind of help he was not taught to provide, and it was impossible for me not to reach back.

Before going to the Childrens' Hospital, I spoke to several neurologists at UCLA who had had experience in dealing with brain tumors in infants. The picture that emerged was bleak but not altogether hopeless. In some cases, the children were eventually able to think, speak, and control bodily movements. A lot depended on the size and precise location of the brain tumor. Even a small chance of restoration could be a lifeline for the parents.

At the hospital, I met with the physician and his wife in the child's room. I introduced myself to little Donna, who stared at me from her crib. She had deep brown eyes and her dark hair was in ringlets. I reached down and took her hand and was encouraged by a slight responsive movement of her fingers. Suddenly, I remembered the delight of my daughters when they were tiny and I would fish dimes and quarters out of their ears.

I took a quarter from my pocket, held it up before Donna, then seemingly tossed it away. The infant's face was expressionless, but her eyes followed my hands.

I asked Donna where the quarter had gone. Her mouth opened as though to say something. I asked if the quarter was behind Mommy's ear. Mommy shook her head. I asked if it was behind Daddy's ear, and he shook his head. Then I said it must be behind Donna's ear, and I reached down, plucked out a quarter, and held it up to everyone's astonishment. Donna's eyes danced with delight. She made a slight gurgling sound.

It was evident that, however severe the surgery, some cognitive function still existed and that, with time, it might be substantially developed. At the very least, mother and father need not be deprived of their hopes. Indeed, those hopes were an essential part of the nurturing process so critically needed by the child.

Outside Donna's room the parents and I chatted. Since no one could say that significant brain development was barred, it was important to hope for the best and to work for the best. I was certain they realized that their own attitudes in the child's presence must not convey apprehension or despair.

I walked to the hospital elevator and passed the large rehabilitation room, brightly lit and multi-colored. Crippled children were moving about, some of them on crutches or in strollers. An armless infant sat on a toy wagon being pulled by a child of eight or nine whose head was heavily bandaged. I looked at the children and the years fell away and I was back at the tuberculosis sanatorium in New Jersey more than a half-century earlier. I remembered visitors looking at us as they walked through the wards. I remembered thinking how lucky they were to be able to come and go. How I wished, looking back, that it might be possible to transmit to each child suffering from illness or disability the sense of what I later came to know was true—that regeneration is the greatest force in life, that it is not foolish to dream of better things, and that we discover ourselves as human beings when we move in the direction of our dreams.

For years after I left the sanatorium I would say nothing about it, fearful that people would shy away from me. Contact with people who have had serious disease, even after they are fully recovered, doesn't come easily in this society. Eventually, however, I no longer dodged the subject. It is important for people to know that disease can be incontestably conquered, and that what is truly risky is the notion that illness never really takes its departure and that we are fated to coexist with a silent and malevolent incubus.

Yes, I know that my optimism can be carried too far—but

I also know that no one is smart enough to fix its limits; and that it is far better to pursue a remote and even unlikely goal than to deprive oneself of the forward motion that goals provide. My life has been an education in these essentials, and I could no more turn away from them than I could deny my own existence. Sharing the convictions that come out of these essentials, of course, is not without risk. Some people may interpret the effort as overwrought do-goodism; but the risk is worth taking nonetheless, for every once in a while it happens that hope can rekindle one's spirits, create remarkable new energies, and set a stage for genuine growth. At least nothing I have learned is less theoretical than the way the entire world seems to open up when courage and determination are connected to truly important aims.

5

THE INFINITE WONDER OF THE HUMAN BRAIN

AND HOW IT AFFECTS EVEN AIDS

Initially, as I've said, my appointment at the School of Medicine was in the Department of Psychiatry and Biobehavioral Sciences, but after a few months, it was extended to include the Brain Research Institute, then headed by Dr. Carmine Clemente. (After I had been at UCLA for several years, the appointment was extended further still, when I was also assigned to the dean's office.)

At first, my office was in Slichter Hall, in a section occupied by the BRI. Shortly after I moved into my new office, Dr. Clemente paid a visit, welcomed me to the faculty of the Brain Research Institute, and told me of its work. He introduced me to the world of the human brain—a region of infinite wonder and splendor.

Dr. Clemente had an impressive and varied background. He was author of a widely used textbook on anatomy. He was also the author of an updated version of the famed *Gray's Anatomy*, a basic textbook for medical schools for a century or more. He came to UCLA in 1952 as an instructor in anatomy in the newly formed School of Medicine, moving up through the ranks to become professor and chairman of the Anatomy Department in 1963, a position he held for ten years. UCLA medical students have been the beneficiaries of his accumulated wisdom and knowledge through thirty-five years of the teaching of anatomy, and Dr. Clemente has been recognized for achievements in anatomical and neurological research by numerous prestigious awards.

For eleven years, beginning in 1976, Dr. Clemente directed the Brain Research Institute at UCLA, a challenge he met with distinction. The institute is located in a ten-story building, housing 135 research laboratories, where 140 UCLA faculty members are involved in neuroscience or brain research.

I learned that the BRI, under Dr. Clemente, had as its central mission the most fascinating task in the world—to train the human brain to scrutinize itself. The BRI was attempting to use human intelligence for the purpose of understanding the source and workings of intelligence.

It was a formidable undertaking. Not even the universe, with its countless billions of galaxies, represents greater wonder or complexity than the human brain. The human brain is a mirror to infinity. There is no limit to its range, scope, or capacity for creative growth. It makes possible new perceptions and new perspectives, just as it clears the way for brighter prospects in human affairs.

If the brain of an average fifty-year-old person could be fully emptied of all the impressions and memories it has stored, and recorded on tape, the length of the tape would reach to the moon and back several times. Indeed, it is possible that the memory contents of the human brain could never be fully inventoried; new impressions would come in faster than the old ones could be identified. Much has been said about the memory capability of computers, but the computer has not been devised that can match the potential capacity of the human brain. Silicon chips and semiconductors have been hailed as the supreme achievement of technology, but these chips are far inferior to the neurons of the human brain in terms of function.

The human brain presides over the biological wonderland that is the human body. The average brain weighs approximately three pounds, accounting for about 2 percent of the body's total weight yet consuming more than 20 percent of its oxygen.

The brain does most of its work through its neurons—agents too small to be weighed or measured. No one knows how many neurons reside in the brain. The estimates move sharply upward decade after decade. As recently as 1950, brain researchers thought they were being extravagant when they guessed there might be as many as a billion neurons. Today, the estimates range from 50 to 100 billion. These neurons carry the traffic for millions of signals. The brain of a writer reaching for just the precise word can set off millions of electrochemical reactions. When we visualize the face of someone we know, the same process occurs. In short, numberless signals are flashing almost every second of our waking hours. Hooking up our thoughts in sequence involves vast neural activity.

The main control console of the brain is the hypothalamus, a vital link between body and brain. Its situation below the cerebral cortex and atop the spine seems most appropriate for this purpose. This is where the basic drives of the body—sex, hunger, awareness of danger, etc.—are located. Similar control is exercised by the hypothalamus over the pituitary gland, which governs the production and circulation of hormones for the body.

The emotions begin with sensory input from the nervous system. Dr. Clemente has eloquently described this transmission of information: "The touch of a hand of a laboratory colleague who is simply a non–emotionally bound colleague produces quite a different type of emotional input than, say, the touch of the hand of a lover. What that means is that nerves all over the body can be involved in transmitting information to the brain—information from the five senses and the organs, information about body movement, and information of an emotional nature. Now it has become evident that many other substances circulating in the blood, besides hormones, make direct contact with brain centers as well."

All this sensory input, which begins in the brain, has its effect through the body. Few aspects of the brain are more fascinating or significant than the way it makes its registrations on the immune system. In this way, our thoughts can have an effect on health and our ability to turn back disease. Brain cells and immune cells are equipped for direct communication with one another.

These connections are so intimate that, according to Dr. Elena A. Korneva of the Institute for Experimental Medicine in Leningrad, USSR, stimulation to the front of the hypothalamus (which is involved in the calming of emotions and the body's ability to absorb nutrients for the healing process) *increases* the body's immune capability, while the back of the hypothalamus (which relates to stress, as in the "fight or flight" response), impairs the performance of disease-fighting cells in the immune system. Dr. Viktor M. Klimenko, a colleague of Dr. Korneva's, has shown that immune responses in turn cause chemical and electrical changes in the brain.

The brain is the only organ of the body that is totally essential for individual identity. If we have a defective kidney or liver, or even heart, we can acquire a transplant and still retain our sense of self—who we are, what we have done or want to do, a knowledge of our commitments and our aspirations. But if we were to acquire a new brain—assuming that medical science could solve the incredibly complex problems involved in a brain transplant—we would also acquire a new self.

At one time, a compartmentalized view of the human body was generally accepted. Recent knowledge of the anatomical and functional links between brain and body point in a different direction. Brain researchers now believe that what happens in the body can affect the brain, and what happens in the brain can affect the body. Hope, purpose, and determination are not merely mental states. They have electrochemical connections that play a large part in the workings of the immune system and, indeed, in the entire economy of the total human organism.

In short, I learned that it is not unscientific to talk about a biology of hope—or of any of the positive emotions. The emotional state of the patient has specific effects on the mechanisms involved in illness and health. The modern physician, therefore, does not confine himself to physical symptoms in diagnosing or treating an illness but probes for possible emotional or stressful factors. He will prescribe not just out of the pharmacy or his little black bag but out of the magnificent apothecary that is the human brain, which can activate and potentiate the healing system. The roster of emotions that figured in the obsession that had brought me to UCLA—hope, faith, love, will to live, festivity, playfulness, purpose, and determination—are powerful biochemical prescriptions.

Meanwhile, we can contemplate the fact that our brain is not being fully used. No one knows what the limits are to the functional capacity of the human brain. It is logical to believe that there are limits to everything, but researchers who have been probing the question report that their studies have yet to yield a clue to the brain's functional limits. The reserve capacity of the brain, therefore, represents the greatest asset of the human species in meeting any problems or challenges, however complex, that the future may hold. The most reassuring fact of life is that the human species is equal to its needs; there is no problem, however bulging or complex, beyond effective human response. What is most significant of all about the human brain perhaps is that it enables the individual to do something that no one has done before. So long as this is so, human beings are the most privileged species on earth.

Ever since scientific investigators have begun to probe the structure and function of the human brain, the main research has been connected to consciousness and cognition. Just in the past half-century, as stated elsewhere, new findings have emphasized the glandular role of the brain. Indeed, new research suggests that the brain may be the most prolific gland in the human body. Some three dozen secretions are produced or activated by the brain. The brain also has the ability to combine these secretions, meaning that there are literally thousands of prescriptions the brain can write to meet the body's needs. Some of these secretions—endorphins and enkephalins—act as the body's own painkillers. Endorphins and enkephalins may play a role in regulating immune function and tumor growth as well. For example, Dr. Nicholas P. Plotnikoff, of the University of Illinois, has discovered that in cancer and in AIDS patients, altering the level of enkephalins determines whether immune responses are enhanced or decreased. Other secretions, such as interferons, which are produced by the brain, as well as by the immune system, have a general immunological function. They were at first thought to have a role that was limited to combating infections. Later research, however, established the fact that interferons not only could attack viruses but could inhibit the growth of certain tumor cells and trigger other immunological defenses against cancer.

The human brain serves as a control post for millions of messengers carrying instructions to the body's organs without intruding on the conscious intelligence. Recent research tends to emphasize "communication" rather than "connection" to describe the process by which the brain interacts with the body. This two-way communication system is as varied, complex, and busy as the operations of any air control tower at the world's largest airports. Directions being given by the brain travel along heavily trafficked electrochemical pathways. These directions are augmented by delivery of specific activating substances.

For example, Dr. Candace B. Pert, of the National Institute of Mental Health, has undertaken research indicating that hormonelike substances called neuropeptides serve as information carriers to coordinate the functions of the brain, glands, and immune system. She refers to neuropeptides as the "biochemicals of the emotions"—or substances that help translate emotions into bodily events. Dr. Pert has shown that the limbic system (the seat of the emotions in the brain) is a focal point of receptors for neuropeptides. The entire lining of the intestines is also richly imbued with neuropeptides and neuropeptide receptors, which suggests to her why a lot of people experience emotions as a "gut feeling." Furthermore, her husband and colleague, Dr. Michael Ruff, has discovered that every neuropeptide receptor—there are at least fifty or sixty in all—is found in cells of the immune system. Immune cells are also capable of producing neuropeptides.

One of the most interesting aspects of the brain's multiple activities is its activation of the body's backup systems. For example, if one part of the respiratory system malfunctions, other parts are directed by the brain to take up the slack. The failure of one lung results in added function by the other. Ditto, the kidneys. The evidence is mounting that the heart can make its own bypass under circumstances of arterial blockage. Narrowing of the heart's arteries can be caused by a diet heavy in fats and cholesterol and by a life-style high in stress and low in exercise. But the elimination of these faults can sometimes produce a natural bypass in which the heart receives an augmented supply of oxygen through a rich new network of blood vessels. Heart attack sufferers who are able to apply themselves diligently to a new life-style stand a good chance of reaping these benefits. The brain is the command post directing these functions.

One of the most fascinating questions to confront medical investigators has to do with the body's backup system in cases of failure of parts of the immune system. The major characteristic of the AIDS disease, for example, is that it knocks out helper T cells. This breach in the body's defenses creates a clear path for a wide range of attacking microorganisms. The prevailing viewpoint is that the helper T failure is not accompanied by heightened activity of other immune cells, but Dr. George F. Solomon of UCLA and his fellow researchers have been developing the evidence that the immune system is not altogether without backup after destruction of helper T cells. In studying AIDS patients who have lived far past the time of original diagnosis, Solomon has documented increased activity of the wide range of immune cells other than those directly attacked by the virus.

Why should this augmented immune activity be in greater evidence with some AIDS patients than others? Here a significant fact comes into view. The AIDS patients who live long past the time predicted for them seem to have a certain trait in common. Perhaps the most important of these characteristics is the refusal to accept the verdict of a grim inevitability. Like the cancer patients at the Wellness Community (described in Chapter 6) who do not deny the diagnosis but who defy the fatal outcome that is supposed to be connected with it, the AIDS patients studied by Solomon have a blazing determination to persevere and prevail. They do not accept the fatalism so characteristic in public thinking about the disease. They provide emotional support for one another. Their personal horizons are uncluttered by defeatism and inevitability.

Can the human brain actually convert attitudes into specific immune system change? Later in this book, we describe a UCLA research project involving cancer patients whose immune systems experienced a sharp boost when they were liberated from the depression that almost invariably follows the diagnosis of a life-threatening disease. The fierce determination so characteristic of these patients cannot be discounted in any evaluation of their condition. Medical science learned from Dr. Walter Cannon a half-century ago that such heightened emotional states can stimulate the spleen with a corresponding increase in the population of red blood cells. Today we know that the spleen has a role in the functioning of the entire immune system.

The effectiveness of brain-spleen-immune interaction came to my mind when a dentist and his lawyer came to visit me at UCLA early in 1988. The dentist said he was going to die of AIDS—he was HIV positive—and he wanted to leave a considerable sum of money to support our research at UCLA. He had brought his lawyer along in order that the terms of the bequest to UCLA could be correctly stated. I told the dentist that I would not talk to him about the bequest but I would be glad to work with him if he wanted to mobilize his will to live and his determination to fight the disease. He accepted the challenge; I put him through some of the mind-body exercises developed by Dr. Elmer Green of the Menninger Foundation clinic, described in Chapter 6.

Three months later, the dentist returned, bearing medical documents showing that he was no longer HIV positive. The look of incredulity on my face produced a wide smile on his.

"My doctor gets the same reaction when he talks about me to other doctors," he said. "Some of his friends say that it can't be, that it was a mistaken diagnosis, but my doctor knows better. He checked and double-checked the original tests and he knows it was no mistake."

I discussed this episode with Dr. Karen Bulloch, immunologist of the University of California at San Diego medical school, who said that five cases of reversals of HIV diagnoses had been reported in the medical press. Dr. Solomon called my attention to a half-dozen AIDS patients in San Francisco who were still alive seven years after diagnosis. Tests of their immune systems showed that there was actually some enhancement of the helper T cells. Among persons with AIDS who were tested at varying times after diagnosis, Solomon and colleagues found some correlations between adaptive emotional attitudes (such as the ability to say no to doom) and better immune measures (including helper T cell numbers). The same correlations appeared in regard to negative emotions, such as dejection, anger, or hostility, and poorer immune measurements.

In any case, Dr. Solomon pointed out, just the fact of an HIV-positive finding did not mean that the disease would abruptly claim its victim. Some people who receive an HIV-positive diagnosis go on for years with the disease in an apparently dormant state. Any attempt to predict, at the time of the diagnosis, those persons who would go into the active stage and those who would not was still sheer guesswork. A multitude of factors was involved, many of which were yet to be identified. The very fact that the illness did not assert itself for a time indicated that the immune system was not altogether nonfunctional. The big hope in AIDS is not just that a specific countermeasure may be discovered or devised but that methods may be found for enhancing the components of the immune system not knocked out by AIDS.

It is important for people to put terror and defeatism behind them in thinking about AIDS. Indeed, one of the main impediments to an effective attack on AIDS is the public hysteria associated with the disease. This hysteria produces a climate in which persons who are diagnosed as HIV positive go into a state of emotional collapse that in itself compromises both treatment and potentiation of the patient's own resources. AIDS is another example of the fact that the way in which we think about a disease has an effect on the outcome.

This is not to say that the answer to AIDS is the avoidance of a defeatist attitude. What it means is that any progress in coping with the disease involves not denial but a vigorous determination to get the most and the best out of whatever is now possible.

6

DENIAL, NO. DEFIANCE, YES.

Within a period of two weeks, two cases that came to my attention were almost identical. The first patient, a young man of twenty-nine, had been given a diagnosis of multiple sclerosis. He spoke Spanish and his wife translated. He came with his young wife and baby of four months. He had received at least a half-dozen different diagnoses over a period of six or seven months for his assorted symptoms but finally received a report of tests that served as the basis for the MS diagnosis—a diagnosis, incidentally, with which later consulting physicians agreed.

The young man said he had visited a chiropractor's office and that the chiropractor had pressed heavily and suddenly on the spine, following which the young man began to develop a pain in his left leg. He said that this was the beginning of other symptoms leading to the diagnosis of multiple sclerosis. What was most striking about the case was that, shortly after the young man received the MS diagnosis, his pains escalated at least threefold.

"It was amazing," he told me. "I had had a pain in my left leg; it was bad but not so bad that I couldn't function. Then the doctor said 'MS' and it seemed as though the pain exploded throughout my entire body. One minute I was existing on one level, difficult but bearable; the next minute I was sick all over."

I looked at the young wife. She was holding her baby and her gaze shifted from her child to her husband. What about her thoughts? Her marriage was less than two years old and she had hardly become a mother when her family was burdened by terrible uncertainties.

It was soon apparent that the young man was in the grip of the deep depression that almost universally follows a serious diagnosis. I referred him to a psychotherapist who had had good results in working with patients in similar circumstances. I also instructed him in the technique of "moving his blood around," borrowing from the program developed by Dr. Elmer Green, mentioned earlier. When the young man was able to see the proof that he could increase the surface temperature of his hands by more than 10 degrees, his feelings of helplessness receded sharply. The fact that he had a measure of control helped him to reduce his depression. What was most important was that he felt he was not consigned to a passive role but was able to work in tandem with his physician.

The second case, as I say, was almost identical. The patient, twenty-six, accompanied by his father, came into my office. He sat down slowly and shook his head.

"It all happened so quickly," he said. "One day I was in charge of my life. I was confident that my pains would lessen. Then the next day I was terribly sick, hardly able to move. My pains were awful, especially in my legs and my back."

"What happened to bring about the sudden change?" I asked.

The young man choked up and his father answered the question.

"What happened was that we were at the specialist's office. He had the results of the tests in front of him and he said my son had multiple sclerosis. Just like that. The roof fell in. By the time we got home I thought my son had aged twenty years. He was walking like an old man. Every step was agony for him. He couldn't get going in the morning. It was like paralysis had hit him."

"Surely," I said, "you've had symptoms for some time and must have known something was wrong."

"They came and went," the young man said. "About a half

year ago, I began to feel weakness in my muscles. After a while I went to see my doctor. He took some tests and said the muscles were okay and that rheumatism was probably the cause. It could have come from toxicity caused by the medications. He said the pains would probably go away, and they did. But after a couple of months the pains came back. They weren't too bad, not as bad as the time before. Nothing that I couldn't handle. My family doctor took me to see a neurologist. He was the one who gave us the bad news. All at once, everything inside me collapsed. I felt so helpless. I didn't feel like doing anything. I wanted to die and get it over with."

What does one do in such a case? The young man hoped he might be helped to think about his illness in a way that allowed for some hope. I referred the young man not only to a psychotherapist, as I had done in the previous case, but to a primary care physician who would be able to quarterback the case. The physician was Dr. Mitchel Covel, a colleague at UCLA, whose proficiency in medicine is accompanied by extraordinary compassion and communicating skills. What was most significant to me was that every patient I referred to Dr. Covel came away reassured. Even when the condition was serious, the patients were reinforced emotionally and able to cope with the psychological problems that accompanied the illness.

The two cases just referred to were not unlike hundreds of patients I have been asked to see during my ten years at UCLA. They suffered from various symptoms, went to see their physicians, were referred to specialists, were put through copious tests, were given a flat-out diagnosis of serious illness, panicked, and experienced a sudden and severe intensification of the underlying disease.

A colleague in the medical school told me about a woman who had gone in for a complete checkup and learned that one of her kidneys was totally nonfunctional. The shock produced instant deafness. With sustained and prolonged psychotherapy over several months her hearing was restored. What troubled my colleague was that the infirmity need not have happened. The routine nature of the way the diagnosis was communicated produced a health problem hardly less severe than the original deficiency.

Most of the patients I am asked to see are cancer sufferers

but almost every serious disease is represented in the total group—multiple sclerosis, lupus, scleroderma, diabetes, heart disease, Parkinson's. What is most striking about these cases is that the illness worsened coincident with the diagnosis.

Why should these patients experience a severe downturn in their condition coincident with the diagnosis? Why should bad news make them worse? Was it possible that, the moment they had a label to attach to their symptoms, the ability of their bodies to respond to a challenge was significantly diminished?

Was I making a reasonable deduction? While I brooded over these questions, I saw a small news item in the *Los Angeles Times* that gavc mc a clue. The item reported an episode occurring at the football stadium in Monterey Park, on the East side of Los Angeles. A handful of persons reported ill during the game with symptoms of food poisoning. Through careful questioning, the examining physician ascertained that all of them had consumed soft drinks from the dispensing machine under the stands. He considered the possibility that the syrup in the drink might have been contaminated. He also wondered whether copper sulfate from the copper piping might have leaked into the drinks. In an effort to protect the spectators, school authorities caused an announcement to be made over the loudspeaker requesting that no one patronize the dispensing machine because some persons had become ill with food poisoning.

The moment this announcement was made, the entire stadium became a sea of retching and fainting people. Many hundreds fled the stands to rush to their physicians or their basins, whichever came first. Many hundreds more were too ill to move and ambulances from five hospitals had to ply back and forth between the stadium and the emergency rooms. More than one hundred persons had to be hospitalized after emergency treatment. Then it was ascertained that the soft drinks were entirely innocent. Whatever the precise cause, at least the dispensing machine was no longer suspect. This news was passed along. The illnesses vanished as suddenly and mysteriously as they had appeared. The patients got up from their hospital beds and went home. The critical agent was language—both in causing illness and overcoming it. Words were processed by the human mind in a way that made for illness or recovery.

Hence the question: If people at a sports event could become sick enough to be hospitalized just as the result of an announcement over a loudspeaker, what is the effect on someone who has actual symptoms, goes to the doctor's office, and hears words like "cancer" or "heart disease" or "diabetes" or "multiple sclerosis" or any of the other life-threatening diseases?

Hardly less striking than the effects of a panic reaction to a serious diagnosis among the cancer patients I saw was the fact that some patients had had an opposite experience. They responded with a fierce determination to overcome. They didn't deny the diagnosis. They denied the verdict that is usually associated with it. Was it any coincidence that a substantial number of these patients lived significantly longer than had been predicted by their oncologists?

Almost two hundred of these survivors have come together in a group known as the Wellness Community. They have a building or community center of their own in Santa Monica, California, provided by Harold Benjamin, an attorney and civic-minded citizen. These people have joined forces not just for the purpose of celebrating their longevity but to help meet the emotional needs of newly diagnosed cancer patients.

I visited the Wellness Community and chatted with some fifty or sixty members of this group. I was eager to learn how they themselves perceived the reasons for their extended longevity. One woman in her upper seventies—she had the kind of serene and majestic beauty one associated with Grace Kelly—said she had a very clear idea of a turning point for the better in her own case. It came, she said, very early. The physician told her he had the results of all the tests at hand and was making an unequivocal diagnosis of cancer. He said it was terminal and gave her four to six months.

"I looked him straight in the eye," she said, "and told him to go straight to hell. God could give me four months to live but not another human being. That was six years ago."

Everyone cheered. Mrs. A. epitomized the experience of the people in the Wellness Community. As I said, they didn't deny the diagnosis; they denied and defied the verdict that was supposed to go with it. The defiance took the form of a blazing determination and it was a window on the future. Their longevity beyond the predicted time ran from a half year to nine years. Most of them are still under treatment. They are able, many of them, to increase their tolerance of chemotherapy by subordinating their fears of it to their awareness that modern medical science is able to combat cancer directly inside their bodies. They do not expect that the battle is without strain or sacrifice. But the important thing is that they were able to summon all their resources for the fight.

The Wellness Community experience emphasized for me more than ever that the way the physician communicates a diagnosis can have a bearing on the outcome of the illness. If the diagnosis is delivered in a way that produces panic, helplessness, and depression, then the body's own resources for fighting back are severely compromised. On the other hand, if the diagnosis results in acceptance of a challenge, however severe, an environment can be created conducive to treatment.

I don't think most physicians are insensitive to the emotional needs and responses of patients in the way they communicate their findings. The economic incentive simply runs the wrong way. As I stated earlier, medical plans do not compensate doctors for talking to patients. An even more basic factor in "cold turkey" communication perhaps, is that physicians are worried about malpractice suits. They have been educated by their lawyers to leave no doubt about anything that might happen during the course of an illness. An unequivocal or flat-out method of communicating a diagnosis is regarded by some physicians as a way of protecting themselves against an expensive lawsuit. But negative expectations can be planted in a patient's mind in a way that can compromise effective treatment.

Insurance premiums against malpractice suits in some areas of the country can cost upward of $100,000 a year. Gradually, however, physicians are learning that the best protection against a malpractice suit is a good relationship with patients. It is not difficult to recognize that the doctors who are sued least are in family practice. The more specialized the branch of medicine, the greater the risk of malpractice suits. Proctologists and gynecologists are among those who find themselves in court more often than internal medicine practitioners.

Compassion, personal interest, and communication artistry are more than just refined and essential skills. They are probably the best insurance policy against lawsuits a doctor can obtain.

They also help to create an environment in which the physician's special abilities can show to greatest advantage.

When I asked my colleagues in the medical school whether I was making unwarranted assumptions about the connection between a panicky response to a diagnosis and a downturn in the disease, I was heartened to hear some of them say that there were specific physiological reasons for the relationship. They pointed to new medical research that demonstrated the disabling effects on the immune system triggered by negative emotional states. The connections between emotional states and biological effects I was searching for no longer seemed far away.

At UCLA's biomedical library I was able to find reports in the medical journals of immune system impairment as the result of such experiences. For example, Drs. Sandra M. Levy and Ronald B. Herberman of the University of Pittsburgh and the Pittsburgh Cancer Institute observed that depressive behavior (fatigue, listlessness, apathy) was associated with diminished natural-killer (NK) cell activity and accelerated tumor spread in breast cancer patients. This finding was made even more significant by their corresponding discovery that NK cell activity was highly indicative of tumor spread. Similarly, Dr. Marvin Stein, of the Mt. Sinai School of Medicine at the City University of New York, and colleagues Steven J. Schleifer and Steven E. Keller of the University of Medicine and Dentistry of New Jersey Medical School, studied a group of widowers after the loss of their wives and found a decrease in T- and B-cell activity as early as two or three weeks after the death of a spouse. To confirm the relationship between severity of depression and suppressed immunity, the investigators conducted a series of studies comparing individuals hospitalized for depression with those not hospitalized and those hospitalized for other reasons. They concluded that the severity of depression was associated with reduction of T and B cells and their activity. This conclusion was strengthened by their observation that relief from depression is paralleled by changes in the immune system.

Drs. Janice Kiecolt-Glaser and Ronald Glaser, of Ohio State University, found that highly depressed nonpsychotic psychiatric in-patients had significantly poorer DNA (genetic) repair in immune cells exposed to irradiation than did less depressed patients; and that both depressed groups fared significantly worse with regard to DNA repair than the psychologically healthy, nonpsychotic group. All group differences were sustained through the final measurement point, five hours after irradiation—a time period in which DNA repair is expected to recover to pre-irradiation levels. This finding suggests to the researchers that emotional stress may contribute to the incidence of cancer by directly causing abnormal cell development or by indirectly diminishing immune surveillance or competence.

By the same mental processes, however, the immune response can be strengthened or restored. Use of relaxation exercises and creative imagination were found to be helpful in a study of cancer patients by Dr. Barry L. Gruber of the Medical Illness Counseling Center in Chevy Chase, Maryland, in collaboration with Dr. Nicholas R. Hall of George Washington University and later of the University of South Florida. For one year, the patients were asked to imagine the forces in their immune systems being fully engaged in a war against the cancer cells.

The investigators found that these exercises had the effect of stimulating lymphocytes and increasing the production of antibodies and interleukin-2 cells, enhancing NK-cell activity and augmenting the effectiveness of the cytotoxic T cells. The pattern of immune changes corresponded to the level of relaxation and imagery. Equally interesting was the fact that the patients clearly showed intense determination to overcome their disease. (Research described in Chapter 16 shows that such determination may be helpful in combating cancer.)

When the body's own responses are mobilized together with the cancer-fighting techniques of modern medical science, the combination not infrequently can bring about remarkable results. But when the patient's own immune system is impaired, an extra burden is put upon the external measures, such as chemotherapy and radiation. These measures work best in combination with the body's own apothecary.

Chemotherapy is not a magical cure. Its justification is that it contains poisons that can kill cancer cells. The severity of the treatment is undertaken on the theory that the gain is greater than the loss. In calculating the gain represented by the damage done to the cancer cells, the physician also has to take into account the loss represented by the damage caused by chemotherapy to the immune system, the body's own method for combating cancer cells and other diseases. The same is true of radiation, which can damage normal tissue even as it kills cancer cells. That is why the wise physician places such heavy emphasis on the need to counteract the severity of chemotherapy and radiation by making the body as strong as possible. A high level of nutrition, freedom from depression and unusual stress, a strong will to live—all these are vital factors in a total strategy for combating serious disease.

This being the case, I have attempted, whenever possible, to bring newly diagnosed patients together with persons who have come through similar illness. In the case of the young man with multiple sclerosis, I arranged for him to meet with several survivors. The specific evidence that recovery is possible acts like a tonic and actually enhances the prospects for effective medical treatment. And later in this book, a research project is described that demonstrates a measurable boost in the immune system when cancer patients learned to free themselves of the depression and panic that almost universally follows a catastrophic diagnosis.

Lifting patients out of their depression is not easy but neither is it beyond reach. A few pages earlier, I referred to techniques developed at the Menninger Foundation clinic in Topeka, Kansas, by Dr. Elmer Green for the purpose of helping to reduce the pain of migraine sufferers. The patients are taught simple biofeedback exercises such as increasing blood flow to the hands. These exercises are also used to reduce high blood pressure.

It seemed to me that the same techniques might be highly useful in freeing patients from the feelings of helplessness that accompany severe illness. When, for example, patients see the evidence that they have the power to increase the surface temperature of their skin by 10 degrees or more, their entire relationship to their bodies is apt to undergo a profound change. Most people believe they have no control over what happens inside their bodies. But the discovery that they possess a certain measure of control can be exhilarating and even therapeutic. They become aware that they are able to do things that they had always supposed were beyond the reach of the conscious intelligence. The fact that they have the power to move their blood around as a matter of will, verified by a skin thermometer, makes a profound impression. They realize that they have something to bring to the encounter; indeed, that the idea of a patient-physician partnership is not just a pretty slogan but a functional necessity and reality.

In adapting Elmer Green's hand-warming techniques to the seriously ill patients I am asked to see, I try to incorporate relaxation techniques. It is useful to have soft lighting. I am lucky enough to have an Eames chair in my office. The patient, eyes closed, can sit back and stretch out, legs on the footstool. The patient is instructed to hold the bulb of a hand thermometer. We begin with deep, slow breathing. After three or four minutes, the patient is ready for the next step. The sequence goes something like this:

Think back on your life. Think of the nicest thing that happened to you—something that made you feel very good at the time. When this memory comes to mind, just raise your little finger.

Usually, it takes no more than a half minute for the memory to be summoned. The little finger goes up. Then:

Now, make this memory as real as possible. So real you can almost taste it. Imagine that you are reliving that experience. Go slowly. Breathe deeply. Try to feel the way you did during that experience. Let everything about that experience give you the same pleasure now that it did then. Let it feel like honey in your veins. Breathe slowly and deeply. You are feeling very good. If the memory makes you laugh or sing out with joy, just do it.

This part of the exercise takes about five minutes. After the patient appears to be in a state not just of relaxation but of openness and contentment, we move on to the heart of the procedure.

Now, let's go exploring. Imagine that you are able to mobilize your consciousness so that it is like the tip of a blackboard pointer as you move from place to place inside your head. Now let that mobilized consciousness come to rest toward the front of your face, just behind your nose. Concentrate on the tip of your nose. Now, imagine the sensation of touching the tip of your nose with your mind.

Good. Now, elevate that mobilized consciousness until it comes to rest just behind your eyes. Bear down at the point. In a little while you will experience a slight pulsing sensation behind your eyes. When you feel this sensation, raise your little finger.

Reaction time here varies from a few seconds to a few minutes; I have yet to meet a patient who, whatever the time interval, does not experience a pulsing sensation, slight or strong.

Now, elevate that mobilized consciousness even higher until it comes to rest just under the scalp in the middle of your head. Concentrate at that point. Concentrate hard. In a little while you will experience a slight tingling sensation. When that happens, raise your little finger again.

This step generally takes longer than the previous one. When the scalp sensation is experienced, we are ready for the final step.

What has happened is that you were able to move your blood around as a matter of will. Medical researchers have demonstrated that we have a measure of control even over our autonomic functions. Our control is much greater than we realize. Now, move your blood into your hands. You can do it. Just visualize your heart pumping your blood up to your shoulders; now across the shoulders; now down your arms, past the elbow, down the forearm, past the wrist, and into your hands. Go through the procedure again.

It is generally possible to tell from the facial expression of the patient whether the procedure is working.

Now, open your eyes and look at the thermometer. What does it say?

Average skin temperature runs from 76 to 82 degrees. Most patients during the procedure have no difficulty in increasing the temperature of their hands by 10 degrees or more. When they discover the proof of their ability to move their blood around, they become wide-eyed. Sometimes they ask: "Did I really do this myself?"

"Yes. If you can do this, what else can you do?"

The patients get the point. They realize they are not altogether helpless and are not barred from a measure of control. Since their illness has produced a feeling of total helplessness, which in turn figured in a resultant depression, this discovery is akin to rejuvenation of spirits and helps enhance the environment of medical treatment. Another benefit of the exercise is that it teaches the patient how to program himself or herself to undergo strenuous medical procedures such as chemotherapy or radiation.

It must be said that not all patients are responsive to these exercises. I was asked to see a ninety-five-year-old man who had prostate cancer. He went through the initial breathing exercises without a hitch, but when we came to the next step and he was asked to think of the nicest thing that ever happened to him, he shook his head.

"Nothing nice ever happened to me," he said.

I tried to conceal my incredulity.

"Perhaps I didn't explain myself clearly," I said. "Just think back on your life and pick out an especially pleasant memory."

"I said nothing nice ever happened to me."

"Excuse me, sir," I protested. "What about your marriage day? Wasn't that a pleasant experience?"

Eyes closed and breathing deeply, he thought for a moment.

"Yes, I thought so at the time. Three days later I changed my mind."

I was losing ground rapidly but didn't want to give up.

"Well," I said, "think back on the first time you picked up your firstborn. Wasn't that a pleasant memory?"

"It was a boy," he replied. "And every time I think of the mess he later made of my business, I feel I should have throttled him in his crib."

I suppose I should have given up at that point. As I looked around the room, however, my eye fell on a citation for community service. It was signed by Eleanor Roosevelt.

"Sir," I said. "I am looking at this citation you have for your community service. Surely you must be very proud of that, especially since it was signed by Eleanor Roosevelt."

He shook his head.

"She was a Democrat," he said.

I knew when I was licked. It seemed to me possible that he had lived a long life not despite his temperament but perhaps because of it. I shook his hand and told the gentleman that there was nothing wrong with his control and that he had a great deal to teach me. I said that the next time I needed any advice or help, I hoped I might come to see him.

Now, for the first time, he grinned. His firm handshake went along with the smile.

The gentleman lived for another six months and died just before his ninety-sixth birthday. I was certain his children, grandchildren, and great-grandchildren were free of any feeling that anything was left undone. I know I had no sense that the gentleman needed outside steering. If he had a problem, it certainly wasn't control.

Not all my visitors had health problems. The most memorable of them was an elderly woman whom I shall call Mrs. Cole. I judged her to be in her early eighties. She was pert, tiny, almost birdlike. Her hair was dyed strawberry pink and was done up in ringlets. She wore a bright white fur cape. When she spoke, some of the words underwent a transformation. For example, "bother" came out "bodder."

"I need your help," she began. "Things have been happening to me, no one would believe. I have a friend, Anna, she's a psychic. She warned me my home was going to be burglarized and please be careful. I asked when this was going to happen, she said the next Wednesday night about two A.M. I told her my little grandchildren were coming to visit us and she said to be sure to have them come another time.

"The strangest thing of all she told me was that it wouldn't do any good to call the police, and when I asked her why, she said the police would tell me, don't bodder."

By this time I wondered what I was getting into, but I had passed the point of no return and had to see it through.

"I didn't want to worry Harry—he's my husband," she continued. "Anyway, he doesn't believe in psychics and he yelled at me once when I told him about Anna. So that Wednesday night, after Harry fell asleep, I wanted to guard the house and I went into the living room, where I could watch things.

"Sometime after midnight, I heard the knob turn in one of the glass doors leading from the living room to the patio. The door opened and a man stepped into the room and started walking to the dining room, where we keep our silver. I snapped on the lamp light.

"I didn't want him to panic and do something violent so I said: 'Excuse me, are you sure you are in the right house?'

"He paid no attention to me and kept walking toward the

dining room. I raised my voice enough so Harry would hear. 'You are in the wrong house,' I said. 'Please go away.'

"That woke up Harry and he came into the living room and saw the burglar. 'What do you want?' he said in a loud voice.

"That scared the burglar and he turned and left. Harry went to the telephone and started to call the police. I told him not to call the police because Anna told me the police would say don't bodder.

"That was when Harry really got mad and shouted at me not to mention Anna's name again as long as he lived. He telephoned the police.

" 'Did the burglar hurt anybody?' the police asked.

" 'No.'

" 'Did he steal anything?'

" 'No.'

" 'Did you get a good look at him?'

" 'No.'

" 'Don't bodder. Go back to bed.'

"My friend Anna really knew what she was talking about. I was very grateful and kept going to her. Then, about three weeks ago, Anna told me to be very careful, they were going to try to take our house away.

"I asked Anna if we were going to have another burglary. She said it was nothing like that. It would all be very legal. They would have the right papers. Right away, she said, I must get a good lawyer. When I asked Anna who would try to take our house away from us, she said I would get a sign and would know in time to save the house.

"Then, last Saturday, I got the sign. When I was at the checkout counter in Ralph's Supermarket I picked up a copy of the *National Enquirer* and there was the picture on the front page of the man who was going to take away our house."

"You actually saw a photograph of the man Anna spoke about?" I asked incredulously.

"Certainly. Here, I saved the copy of the paper."

She reached into her large handbag, unfolded a copy of the *National Enquirer* and handed it to me.

"Excuse me," I said. "This is a photograph of Prince Charles."

"Ah," she said. "You think it is Prince Charles. The queen

thinks it is Prince Charles. The *National Enquirer* thinks it is Prince Charles. But it is Georgie—Georgie Whiteman."

"Who is Georgie Whiteman?"

"He used to be a movie producer. I knew Georgie many years ago. Back in those days I was a movie actress. I was the little immigrant girl in Cinderella stories. Georgie wanted to marry me, but I had already told Harry I would marry him. Georgie tried to change my mind. He had been very kind to us. When my mother wanted to buy a house, we didn't have enough money and Georgie loaned us the money, which he never asked us to pay back. He was angry with me because I married Harry. He never forgave me.

"He still has the note on the loan. From what Anna tells me, he wants to foreclose. I keep thinking of the last time we spoke years ago when he said if I didn't marry him I would be very sorry one day. He's going to take our house away, all very legal."

I looked at my watch. The saga of Mrs. Cole had taken almost an hour.

"Mrs. Cole," I asked, "what would you like me to do?"

She looked at me entreatingly.

"I need a lawyer who won't charge an arm and a leg."

Fortunately, my secretary interrupted just then to say I was overdue for an appointment outside the office. That ended the meeting. Later, I wondered what I would have thought back at the *Saturday Review* if someone had told me that there would come a time when I would be in the role of ombudsman in California and that I would be asked for help in situations running all the way from complaints about doctor's bills to dreaded events foretold by psychics. Obviously, I wouldn't have believed it, not even if the person who predicted it was named Anna.

Since I go through the hand-warming technique on myself whenever I lead patients through the exercise, I have been able to make good use of the process in meeting certain problems of my own. For example, I had a hard fall on the tennis court in July 1985, resulting in a fracture of the radial bone in my elbow. I went to the emergency room of the UCLA hospital, where the arm was put in a cast and a sling.

A few nights later, I was at a concert at the Hollywood Bowl.

Sitting in the next box was Dr. Frank Jobe, an orthopedist who has attracted national attention because of the famous sports figures he has treated. He inquired about the reason for the sling, then told me not to expect to get back on the tennis court very soon, nor to worry if I didn't have pain-free finger movement. The reason I had to be patient, he said, was that the blood supply to the elbow is very thin; hence full recovery was to be measured in months rather than days.

When I returned from the concert that night I got to thinking about Dr. Jobe's comments. Since the blood supply to the elbow was thin, it seemed reasonable to suppose that the healing process might be speeded up if that supply could be increased. I was fairly adept in increasing the blood flow to my hands. It was a simple exercise of the anatomical imagination to recognize that the blood had to pass the elbow on the way down; I thought I might be able to get the blood to linger awhile in the vicinity of the elbow. I practiced the exercise three or four times and, to my great delight, discovered increased movement of my fingers. I repeated the exercise several times each day for the next week, with progressively good results. The cast and sling came off about ten days later and I was able to get back on the tennis court the following week.

Two years later, while playing tennis doubles with my wife, we collided while running hard to reach a low ball in the center of the court. I vaulted over her in order to avoid landing on top of her, falling on my side. She banged her chin on the hard surface, biting through her lip. She was treated by her dentist. I went to Dr. Jobe's treatment center, where X rays revealed a broken rib. I was fitted out with a rib belt and instructed how to lie down, stand up, and do all the things we take for granted. Once again, I was admonished to be patient.

Mindful of my experience with the fractured elbow, I naturally applied the same technique to my ribs. The speedup in recovery was remarkable. It was only three weeks before I was playing tennis and golf again.

Command over one's autonomic processes is of course common with yogis. Some years ago, while in India, I visited one of the nation's leading training centers for yogis. There I witnessed techniques for slowing down the pulsebeat. Yogis can perform other feats based on a measure of control over the autonomic nervous system. The hand-warming exercise involves the same mind-body interaction. It demonstrates the mind's ability to exert conscious control over various functions generally believed to be locked into the involuntary functions. It is not particularly difficult, for example, to control blood pressure within a certain range. When Dr. David Cannom takes my blood pressure, he is entertained by my ability to increase or decrease systolic pressure as a matter of will, sometimes by 20 points or more. At first, he thought that I was synthesizing anger, resulting in a measurable blood pressure increase. Then I demonstrated to him that the rise in blood pressure was the direct result of conscious direction of blood flow to my hands or my head.

What we are discussing here is not some arcane Oriental practice or a mystical experience but Elmer Green's procedure. It has salutary applications, one of which, as mentioned a moment ago, can be useful in relieving feelings of helplessness and depression during serious illness. True, biofeedback can be distorted into a vaudeville show, but in its serious use it can be part of the complex equation of effective treatment. Like hypnotic therapy, biofeedback is susceptible to abuse but it can be equally helpful in enlarging the patient's sense of his or her powers.

One of the unfortunate aspects of health education is that it tends to make us more aware of our weaknesses than of our strengths. By focusing our attention and concerns on things that can go wrong, we tend to develop a one-sided view of the human body, regarding it as a ready receiver for all sorts of illnesses. The most important health lesson of all to be learned is that the human body is a beautifully robust mechanism, capable of attending to most of its needs. But we know very little about the body's own processes for dealing with such needs. We have allowed pain to intimidate us unduly. We make the mistake of equating pain with disease; very little is said about the fact that most pain belongs to a warning system to tell us that we are doing something wrong. We may be eating too much or eating the wrong things; we may be smoking too much or drinking too much; we may have more emotional tension or congestion in our lives than we can readily handle. There may be problems within the family or at the office that are increasingly beyond control.

Instead of understanding the message that pain is trying to give us, and attending to the cause, we reach almost automatically for this or that painkiller. Again I emphasize that we don't seem to know that we can have pain without disease, or that it is possible to dispatch the pain without pills. We have become a pill-popping and self-medicating society; in fact, we are becoming a nation of weaklings and hypochondriacs, intimidated by the slightest pain and prepared to believe the worst. The trouble about believing the worst, of course, is that it has a tendency to invite the worst.

Proper health education should begin with an awareness of the magnificent resources built into the human system. We need to know that we possess mechanisms for warding off disease and for combating it. We need to become aware of the marvelous array of cells that circulate throughout the body, detecting the presence of invaders, reporting their presence and locations to a command post in the brain that has the ability to activate the responsive forces—cells that go directly to the site of the invasion, employing the body's own chemical system for combating infectious agents or correcting abnormal growths. This beautiful system can be impaired by all the self-medication we take in the mistaken notion that this is the only way we are going to subdue our pains. By treating the effects, rather than the cause, we succeed only in intensifying the underlying problem.

Few things are more essential for the national future than the need for Americans to be reeducated about health: education about internal and external mechanisms for warding off disease or coping with it, should it occur; education in the requirements of good health; education that can teach us that panic and defeat are the great multipliers of illness; education about the importance of confidence in repair, restoration, recovery, regeneration; education in the need for a partnership between patient and physician; education in what is meant by the human healing system and how it works best; education in the value of putting our best effort toward maximizing what is possible; and, finally, education that can instruct us that what goes on in the mind can promote or retard health. It is in this sense that head comes first.

7

FALSE HOPES VS. FALSE FEARS

The problems of medical malpractice suits are not confined to skyrocketing insurance costs or other legal liabilities. Perhaps the most serious aspect of malpractice suits is the way they reflect a growing conflict of interests between physicians and patients.

Few things are more delicate or important in dealing with serious illnesses than the psychological environment in which a patient is treated. If the physician communicates a serious diagnosis in a way that produces feelings of despair in the patient, the result can sometimes have a negative effect on the course of the disease. As described in the previous chapter, scientific evidence now exists proving that depression, a common result of a serious diagnosis, can actually compromise the effectiveness of disease-fighting cells in the immune system. The artistry of the physician in communicating a diagnosis, therefore, can help get the most out of medical treatment.

Unfortunately, doctors are being indoctrinated by their lawyers against leaving any doubts about downside possibilities. If there is a possibility that anything can go wrong, the lawyers say the patient should not be surprised. This has led some physicians to tell their patients the worst in order to protect themselves against the possibility of a lawsuit. But telling the patient the worst can sometimes bring on the worst. As we have seen, panic and fear are powerful intensifiers of illness. A patient who leaves the doctor's office in a state of emotional devastation not infrequently experiences negative physiological changes. People tend to move along the path of their expectations.

Predictions about negative outcomes are not infallible. When doctors compare notes among themselves, they discuss remissions or recoveries in cases where all the signs ran the other way. Consider, for example, the recent case of a Los Angeles woman who developed a chest cough. Her physician put her through a battery of tests, including comprehensive X rays, on the basis of which he informed her that she had terminal cancer. Without being asked, he informed the patient that she had about sixty days to live. The immediate effect was catastrophic. The patient, who had been able to carry on with her job despite her illness, went into an emotional and physical slide. She became deeply depressed, virtually stopped eating, and experienced a weight drop in little more than a week of eighteen pounds.

At that point, she consulted another physician, who told her that medical journals regularly carry reports of cancer patients who recovered altogether or inexplicably lived much longer than their doctors had predicted. In view of these remissions, he said he had learned never to speculate about deadlines but to concentrate instead on providing the best treatment that modern science had to offer, at the same time bolstering the patient's will to live and, in fact, getting the most out of *all* the patient's physical, spiritual, and emotional resources. Another thing the physician did was to introduce his patient to someone who had recovered from a similar condition. The woman experienced dramatic improvement. She is still coping with cancer, but the longevity factor has been vastly improved.

Words can be lethal. If a physician is going to do everything within his power to treat a serious case, he will not compromise or complicate treatment by creating an environment of defeat and fatalism. No more important lesson can be learned by medical students than that no one knows enough to make a pronouncement of doom. If a physician decides to treat a case

THE MEDICAL USES OF HOPE

As I was eating breakfast one morning I overheard two oncologists discussing the papers they were to present that day at the national meeting of the American Society of Clinical Oncology. One was complaining bitterly:

"You know, Bob, I just don't understand it. We used the same drugs, the same dosage, the same schedule, and the same entry criteria. Yet I got a 22 percent response rate and you got a 74 percent. That's unheard-of for metastatic lung cancer. How do you do it?"

"We're both using Etoposide, Platinol, Oncovin, and hydroxyurea. You call yours EPOH. I tell my patients I'm giving them HOPE. Sure, I tell them this is experimental, and we go over the long list of side effects together. But I emphasize that we have a chance. As dismal as the statistics are for non–small cell, there are always a few percent who do really well."

—WILLIAM M. BUCHHOLZ, M.D.
The Western Journal of Medicine

despite his apprehensions over a patient's chances, that same reasoning should dictate the need to encourage the patient's will to live. The wise physician pays attention not just to his prescription pad but to the climate of treatment. However remote a remission or recovery in a specific case may appear to be, the physician is obligated to get the most out of whatever may be possible.

How, therefore, is the physician to resolve the dilemma represented by the need to try to reduce his vulnerability to lawsuits while at the same time respecting psychological factors in communicating a serious diagnosis? The answer is that the wise physician doesn't talk about a fatal outcome but about challenge. It is a challenge to both doctor and patient in which they work together as partners. If the physician is concerned about fostering false hope, he should be no less concerned about fostering false fears or false despair—especially where his own fear of legal reprisals figures in the equation.

In previous chapters, we discussed problems caused by inartistic communication of a diagnosis, as in the case of the young men who went into a tailspin after receiving a diagnosis of multiple sclerosis. There are also problems caused by lack of discretion, however unintentional.

A California physician told me about the emotional devastation experienced by his son, seventeen, following surgery for cancer. The day following the operation, the surgeon came into the recovery room and, in thc presence of the patient, told the father that he should expect the death of his son in a matter of days, perhaps a week at most.

The father was shocked not just by the catastrophic news but by the fact that the surgeon had no hesitation in delivering the verdict in the presence of the patient.

"I followed the surgeon out of the room," he told me, "and, as a fellow physician, berated him for his reprehensible conduct. He seemed surprised by my anger and defended himself by saying that doctors had to be honest and that patients should not be deceived. He missed the point. He should have consulted with me first and we could have decided on what ought to be done or said under the circumstances.

"I came back to the room and told my son that I had just chewed out the surgeon and that I had known too many patients who made surprising comebacks to justify the kind of verdict the surgeon delivered. I told my son to disregard what the surgeon said and that we would work together in proving him wrong. My son believed me. He sailed through the first week after the surgery and has been in remission ever since. That was four years ago and my son is living a normal life in every way.

"I suppose doctors feel they are only doing their duty when they level with patients," he continued, "and most of the time their predictions are correct. But even if they are wrong only ten percent of the time, they're taking an awful chance of hurting a patient. Anyway, I thought the surgeon actually went out of his way to hang black crepe over my son."

The patient was fortunate that his father was a physician and in a position to try to counteract the effects of the surgeon's blunder. Not all patients have access to such a ready resource. I have in mind a young Iowa woman, who, like the physician's son, had undergone surgery for cancer and was brought to the recovery room. Her parents went to the recovery room, where, after a brief period, they were joined by the surgeon.

The young woman was apparently still asleep at the time.

"We did the best we could," the surgeon told the parents, "but we couldn't get it all out. I suggest that you take her home and make your home a hospice."

After the surgeon left, the parents became aware that their daughter was sobbing. She had overheard the surgeon. The distraught parents had no way of reassuring her. The young woman lost all will to live and resisted food and medical treatment. Devastated by the pronouncement of the surgeon, she wanted only to die as soon as possible. I received a call from the family physician asking me to come to Iowa to see whether the young woman's will to live could be restored.

The girl's father met me at the Des Moines airport. On the automobile drive to their home, about an hour away, the father told me that his daughter, in her early twenties, was only recently married. She had been in good health but had developed a cough. X rays showed a small lung tumor that turned out to be malignant. Other tests showed liver involvement. Husband and parents knew they were bucking terrible odds. Even if the best that could be done was to offset the baleful effects of the surgeon's remarks and restore her will to live, they would feel the effort was justified. It was important, they said, to improve her quality of life.

The conversation with the girl's father made me think of the cancer-stricken judge whose case I described in Chapter 2. Common to both cases was the devastating effect of the patient's fatalism on the family.

At the house, the husband took me to the bedroom. The young woman had been informed that I was coming and held out her hand. I sat by the bedside and thanked her for allowing me to visit with her.

"You know the situation," she said, speaking in little more than a whisper. "The surgeon doesn't think I have a chance. My family wants me to go on fighting. I'm in great pain."

I told the young woman that I could understand her feelings, just as I could understand the feelings of the parents, but that it was important for her to know that the predictions of doctors or surgeons weren't always right. I told her about the seventeen-year-old boy who underwent surgery and overheard the surgeon say he had only a few days left. That was four years ago.

"Do you think I have a chance?"

"I am not a doctor, but I know many fine doctors and they are reluctant to make downside predictions because they have seen too many patients overcome terrible odds. Besides, they don't want their pronouncements to be a hex. They also know that their treatments work best when the patient's internal medicines are put to work."

I then described the way a strong will to live helps the body's own apothecary to go to work, and I spoke about certain cells in the immune system that carry a natural chemotherapy, prying open the cancer cells and killing them off one by one. But this system seems to work best under circumstances of high resolve. I told her of patients who had refuted grim predictions and had managed to prolong their lives, and that I would have some of them telephone her to talk about their own cases. But there was something else she might want to take into account. And that was the effect on her husband and parents. They wanted to do everything possible for her. They were getting the best medical attention possible. But they also feared that if she accepted defeat, even the best medical attention in the world would fall short. I thought she owed it to herself and to them to do everything she could to meet the challenge. There were no guarantees, of course, but I hoped she felt it was worth the effort—for herself and her family.

She said nothing, but she smiled.

Back in California, I received a telephone call from the young husband the next day. He said his wife was eating again and that she had told him and her parents that she wanted to live. He said the whole world had opened up for them again.

Unfortunately, there is no happy ending to this story, any more than there is to the numberless cases of people who put the best they have against catastrophic illness, only to fall short of their expectations; but there is something about making a supreme effort in coping that gives meaning to life, even under the most trying and poignant circumstances. This holds true not just for the patient but for members of the family who desperately need evidence of a total response to serious challenge.

The young woman in Iowa lived for several months. There were days when she was relatively pain free and felt strong enough to get out of bed—days that brought joy to her husband and parents. Then there came the day when, despite her own best efforts and those of the doctors, the disease made its final claim. Her father telephoned to tell of her passing but he said something I will never forget.

"Mary raised our spirits and helped us go on with our own lives," he said. "What a great human being she was. She gave us a great deal to go on."

8

ILLNESS AND GUILT

The most insistent and critical question put to me, generally by physicians, since the original publication of *Anatomy of an Illness*, runs along these lines:

> Aren't you afraid that the account of your experience in combating illness will create false hopes in people confronting similar serious challenges? What about patients who try to apply cheerfulness, good humor, and a strong will to live—but whose underlying condition does not admit recovery or even improvement? Isn't life difficult enough for such patients without adding a burden of guilt?

Obviously, patients must not be deceived into thinking nothing more is required to combat a serious disease than a strong will to live and a good laugh. But this is not really the main issue. The main issue is that people who are fighting for survival want to be sure that nothing is being missed, that all the resources of medical science are at their disposal. It is in this context that patients also want to be sure that their own resources are fully mobilized.

If the question about giving hope to catastrophically ill patients has validity, should it not also apply to hope connected to medical and surgical treatment? Since people reach out for medical care in hope of recovery, will they feel guilty if their bodies fail to respond to the physician's treatment? Is a physician justified in withholding such treatment because the hopes of the patient may exceed the realities of the case?

The answer, of course, is that nothing should be held back that offers a chance, however remote, of improvement or prolongation—even if complete recovery seems impossible. This applies not just to medical or surgical help but to psychological and spiritual care. Since there are imponderables that elude scientific prediction, the patient is entitled to a full mobilization of resources, including his own.

My own experiences in dealing with patients may or may not be representative; but in the past ten years I have met with many hundreds of patients involved in a life-or-death struggle, and I have not encountered anyone who felt guilty because hope and a strong will to live were not enough to pull him back from the brink. The key question was whether anything was being missed that might have made a difference in the outcome. A sense of defeat is more readily connected to failure to obtain the best in medical care than to feelings of guilt because one's hopes were not enough to carry one through.

Similar values exist with respect to the family. Again, the overarching need is to be assured that the finest medical attention is being provided—and this pertains not just to medical care but to the patient's *own* resources, beginning with a strong will to live and hope.

This doesn't mean that I have not known patients who didn't suffer from guilt feelings. Curiously, such episodes occurred more under circumstances of recovery rather than failure.

A young mother of three children, ages four, six, and nine, was stricken with cancer, the major site being the lungs. The woman didn't respond to treatment and the physician told her husband to expect the worst. The husband's mother flew from Michigan to California to help care for the children and to maintain the home. At first, the children were shielded from the bad news but, as their mother's condition continued to deteriorate, they were prepared for the worst.

For two months, the mother lingered near death in the hospital. Then, in response to heroic treatment, the tumors not only stopped their spread but began to shrink. Her condition improved rapidly and she was released from the hospital. The prognosis was guardedly optimistic.

When the woman returned home, she faced a new ordeal. The children, having already accepted the inevitability of her death, seemed puzzled by her reappearance at home.

"Mommy," the four-year-old asked, "Grandma said you were going to die. Why didn't you die?"

The woman's mother-in-law, having closed down her home in Michigan in order to assume a new function in life—taking care of her son and his children—didn't seem overwhelmed with joy by the new developments. The husband, too, wasn't entirely convincing in word or manner about his relief over his wife's recovery. The neighbors seemed to feel they had expended their sympathy and sorrow without adequate cause. The woman slid into deep depression and it was only a matter of time before the malignant tumors began to reappear. This time the aggressive chemotherapy did not work but in fact contributed to the rapid downturn. Within two weeks of returning to the hospital the woman died.

Just the fact of medical care raises hope. The medical profession would cease to exist were it not for the hopes of patients. The wise physician hesitates to squelch hope. He knows that the emotional needs of patients can affect the course of the disease. Moreover, one of the prime advantages of developing the patient's emotional and intellectual resources is that it sets a stage for the physician to do his best.

The ability of patients to free themselves of the depression and despair that almost universally accompany a serious diagnosis; the ability to have confidence in their physicians and in themselves; the ability to maintain a good quality of life even under the most difficult circumstances—all these responses to the challenge of disease go along with effective medical care.

One reason medical journals regularly carry accounts of remarkable remissions is that they were unexpected by the attending physicians. So long as the smallest chance of recovery or improvement exists, the wise physician will do nothing to dampen the patient's prospects or hopes. Even when the outcome of a case is dubious, the prolongation of life is often a triumph in itself.

The same is true of the patient's relationship to the psychologist or minister or friend who offers encouragement or tries to create a positive atmosphere. Thinking back upon the many hundreds of patients I have seen at the request of their physicians for morale-building purposes, I am struck with the fact that even when the illness continued to worsen, the patient's improved quality of life made an essentially melancholy event more bearable—a change of value to the family as well.

We must not overlook the profound significance attached by many patients to an extension of life even by a few years or sometimes even a few months. To a patient who has just suffered a major heart attack, the realization that death may be deferred for some months can be deeply meaningful. It is not uncommon for seriously ill patients, when talking among themselves, to recognize a hierarchy of longevity. Those who have been informed by their physicians that they probably have only three months to live are envious of those who have six months; and those who have six months are envious of those who have a year. And so it goes. Patients set goals for themselves: They want to be able to be with their families for one more Christmas, or one more wedding anniversary; or be at the graduation exercises of a child or grandchild. Attaining these goals can be a great victory—even though a grim outcome is only being postponed.

Life is the ultimate prize and it takes on ultimate value when suddenly we discover how tentative and fragile it can be. The essential art of living is to recognize and savor its preciousness when it is free of imminent threat or jeopardy.

Here I return to an earlier point. Concern over "false hopes" tends to ignore the far more prevalent problem caused by "false fears." These fears can be caused by the insensitivity and bluntness with which a diagnosis is delivered, a manner that can actually complicate an illness.

What should the sensitive diagnostician do—lie? Certainly not. It is possible to communicate the truth so that the patient is not left in a condition of emotional devastation, itself a risky business. Consider this example of effective but sensitive communication: "We have something serious to talk about, and one of the first things I am going to do is to introduce you to a person who also had cancer and has come through it—even though the statistics pointed the other way. I have tried to learn from this case and others like it. What I have learned is that the full resources not just of modern medicine but of the patient himself have been utilized. Each human being possesses a beautiful system for fighting disease. This system provides the body with cancer-fighting cells—cells that can crush cancer cells or poison them one by one with the body's own chemotherapy. This system works better when the patient is relatively free of depression, which is what a strong will to live and a blazing determination can help to do. When we add these inner resources to the resources of medical science, we're reaching out for the best. And that's what I propose to do—to help you reach out for the best because I believe you are entitled to the best that is possible. I propose a partnership."

The physician knows that his little black bag can carry him only so far and that the body's own healing system is the main resource. He makes no promise of a cure. Since he intends to treat the patient despite the morose nature of his prognosis, he feels obligated to create an optimal environment for that purpose. He has not provided false hopes—nor has he inflicted false fears. He will do his best and help his patient to do the same. As Dr. Bernie S. Siegel counsels in *Love, Medicine & Miracles*, "Getting well is not the only goal. Even more important is learning to live without fear, to be at peace with life, and ultimately death."

Not infrequently, feelings of guilt are connected to the belief that illness is a form of punishment for misdeeds, even though the misdeeds may not always be identifiable.

Such reactions sometimes have their origins in religious belief. Ideas of punishment or rewards are not confined to the hereafter. The concept of a Deity sitting in judgment on human behavior is common to cultures in almost every age and place. Children are admonished that nothing is more certain than divine retribution. The power of Divinity—not just to discern evildoing but to mete out swift and certain punishment—has been a standard instrument of authority throughout history. The literature of almost every culture abounds with instances in which misfortune or illness was described as God's way of dealing with sin or evil.

Modern psychologists and psychiatrists are not baffled by such events. They recognize that awareness of wrongdoing often produces prolonged feelings of remorse and self-condemnation that, lacking catharsis, can actually have damaging effects on the bodily systems and open the gates to disease. Among the patients I have been asked to see for morale-building purposes, I can think of a dozen or more who felt that their illness was punishment for things done or undone. One patient who suffered from assorted gastrointestinal disorders, including ulcers and diverticulitis, couldn't expunge feelings of guilt for having placed her eighty-six-year-old mother in a rest home. On a visit to the rest home one month before the mother's death, the daughter found the mother confined to bed in a modified straitjacket that denied her the use of her arms. The physician explained that she was being treated for a rash and that the retainer was mandated in order to prevent the woman from ripping off large patches of skin to get at the itching. The explanation seemed reasonable enough but the memory of her mother begging and weeping to be released from the retainer, especially under the circumstances of the fiendish itching, haunted the daughter and produced feelings of tormenting guilt. Those feelings were not lessened when the daughter later learned that anesthetic creams could probably have controlled the itching.

After the mother's death, the family doctor arranged for the daughter to see a psychiatrist for the purpose of eradicating the pulverizing feelings of guilt. Even after six months of such expert care, however, the self-torment continued. When the woman asked me how I would feel if I had failed my mother under similar circumstances, I said I would probably be afflicted by the same saturating guilt. But I said that at some point I would have to ask myself whether my mother would have forgiven me. Since I believed profoundly that my mother would be terribly anguished at my remorse, there would have been no doubt in my mind of her forgiveness. Yes, I would have been able to come to terms with my failure, if failure it be, and to have gone on with my life as she would have wished me to do. Moreover, I would have been consoled by the fact that at eighty-six she had lived a full life.

The woman was taking in everything I said, then looked at me searchingly.

"My mother was a wonderful woman," she said softly. "Yes, I believe she would have forgiven me. She was always concerned about the feelings of the people she loved. Maybe I have been too self-centered about this, thinking mainly of my own feelings, instead of how mother would have me feel."

I told the woman that I was confident that she could come to terms with her problem but it was more important for her to aid the healing process by taking advantage of the substantial advancement in the treatment of ulcers and diverticulitis that her physician could provide—once the underlying problem was addressed. Meanwhile, it wouldn't hurt for her to throw herself into rewarding activities.

What do you say to a patient suffering from a high-risk disease when that patient expresses the belief that God has punished her for her lack of attention to a dying sister a year earlier?

It is not useful, I believe, to argue that the patient is mistaken or that she was far more attentive to her sister than she realizes. All this does is to energize the patient to mobilize the counterarguments with specific details. What one can do is to listen carefully and painstakingly and to say that you can understand why the patient feels as she does. One can also say it is entirely natural to experience remorse as the result of feeling that one has not done enough for a loved one. One can add that nothing is more inexplicable in this world than serious illness; but that if cancer were God's way of punishing the wrongdoer, how would one explain the fact that many thousands of young children are afflicted with catastrophic illnesses? Are we to say that these children are being punished for their sins? Even if it is said that the children are being made to pay the price for wrongdoing by their parents, the concept of a vindictive Deity is a terrifying assumption and a terrible burden to have to carry through life.

One can understand that persons hit by catastrophic illness would ask, "Why me?" One can also understand, in the absence of plausible reasons for an illness, that someone might feel he is being punished by a higher force. But it may help a patient who is consumed by guilt or resentment, or obsessed with the idea of divine punishment, to dwell instead on the concept of divine forgiveness. Certainly nothing is more characteristic of the Deity than forgiveness. No religion could exist without it.

One dictionary definition of the verb *to forgive* is to "acknowledge that no debt exists." This acknowledgment goes beyond forgetting or pardoning, both of which imply that debt exists but has been dealt with in one way or another.

I doubt that I have learned as much in life as a reasonably long lifetime should be expected to provide (I write this in my seventy-fourth year), but a few things stand out. I have learned that life is an adventure in forgiveness. Nothing clutters the soul more than remorse, resentment, recrimination. Negative feelings occupy a fearsome amount of space in the mind, blocking our perceptions, our prospects, our pleasures. Forgiveness is a gift we need to give not only to others but to ourselves, freeing us from self-punishment and enabling us to see a wider horizon in life than is possible under circumstances of guilt or grudge.

There are times when we may feel wronged, betrayed, deceived, humiliated. It would be unhealthy not to react against the outrage. But limits need to be set to the emotional punishment such resentments and anger, however justified, can inflict on us. Certainly we ought not grant others the right to give us ulcers. Forgetfulness can be an asset in such cases. Forgetfulness is generally regarded as a defect. But forgetfulness allied to forgiveness is a way of erasing the smudges in the mind that come from prolonged brooding over taunts or insults or injustices, real or imagined. Among the prime assets of the human mind is the ability to cut loose from vengeful or burdensome memories. The easiest way to deepen a grievance is to cling to it. The surest way to intensify an illness is to blame oneself or the Deity.

9

PROBLEMS BEYOND THE DOCTOR'S REACH

With each passing year at UCLA, my office hours became increasingly filled with patients whose medical problems transcended medical treatment. Many of these patients suffered from physical pains or symptoms of one sort or another but their physicians were limited in what they could do because the cause was generally beyond their reach. It became clear to me that physicians are frequently confronted by things they cannot fix—and I refer not to the illness itself but to underlying problems. I would be asked to see these patients not for medical but for morale-building purposes.

Example: A thirty-six-year-old woman suffered from assorted symptoms—occasional chest pains, irregular heartbeat, shortness of breath. Lower back pain and inflammation of the joints complicated the general picture. I learned that her husband had been murdered eight years earlier at the family Thanksgiving dinner. The husband's twin brother was present. During the dinner a man broke into the house, rushed into the dining room, and emptied his revolver at the woman's husband. The assailant was captured. It developed he was a disgruntled former employee of the murdered man's twin brother. The woman's husband had been shot and killed by mistake.

When she was due to appear in court, the woman received threats against her own life. She defied the threats and identified the assailant, who was found guilty and sentenced to a long prison term. Eight years later, however, the man was up for parole. The woman developed severe anxiety about the release. She was panic-stricken by fear of reprisal, especially in view of the assailant's earlier threats.

Obviously, some medications existed to alleviate the symptoms but so long as she was fearful, the medical approaches could go only so far. The woman's lawyer had informed her that her attempt to protest the release of the assailant on parole would have meaning only if there were current threats against her life. Shortly after the release on parole of the assailant, the woman developed a breast tumor, subsequently removed by a lumpectomy. As she had feared, the woman began to received abusive phone calls that were eventually traced to the assailant, whose parole was then canceled. As soon as the deep apprehension was removed, the woman's symptoms began to disappear.

Example: A stockbroker in the San Francisco office of a nationally prominent investment house, who had been employed for more than ten years by the same firm, lost his job in the stock market collapse of October 1987. The broker, forty-four, made the rounds of other investment houses—without success. He became increasingly apprehensive, slid into depression, then began to develop symptoms: blurred vision, pain behind the eyes, headaches, occasional dizziness, heavy feeling in the thighs. His daughter brought him into the hospital for a full workup, including a CAT scan. No neurological abnormalities were found that could account for the symptoms. Nor did the various blood tests reveal possible causes. Even as the tests were conducted, the symptoms intensified, the result in part, no doubt, of the fact that the patient's medical expenses exceeded the insurance limits.

The broker tried other avenues of employment, finally getting a job as host at a newly opened restaurant in San Francisco. The restaurant closed down after only a month and the man was out of work again. When his reserves ran out, he became despondent and then suicidal. He was put back in the hospital with a sharp intensification of symptoms. Family and friends rallied around him, meeting the hospital bills. I was able to arrange some appointments with employment executives of nationally known brokerage houses. At this writing, the appointments have not yet taken place. But just the prospect of being interviewed has improved his general situation. If those appointments should not result in employment, however, I don't know what anyone can do to stave off a serious downturn.

Physicians are called upon to treat this patient and others like him, but the only prescription that would work would be a job, which the doctors may not be able to provide.

"We live with these tragedies constantly," Dr. Kenneth Shine told me. "Some doctors try to duck such cases—and you can't blame them. All they can do is to try to ease the pain, but that's not good enough."

Economic ills and physical ills are closely related. The doctor is at the end of the line, trying to catch people from falling over an edge toward which they are pushed by circumstances beyond the physician's control.

Example: A woman I was asked by her physician to see had had an ovarian tumor successfully removed by surgery. However, her physician feared that unless her will to live were bolstered, it would be only a matter of time before serious illness, whether cancer or another disease, recurred. He told me that she had experienced a number of hammer blows in rapid succession. She was forty-one, married, and the mother of a girl of nineteen. Her husband was a successful electrical engineer. Several years earlier, her niece, fifteen, had come to stay with them while the girl's mother was recuperating from a series of respiratory illnesses. Then one day the niece revealed she had been sexually molested a half-dozen times by her uncle. The physician said that after each molestation, which generally happened in the middle of the night, the uncle would threaten the girl with physical harm if she revealed to her aunt or anyone else what had happened.

The physician also disclosed that the daughter broke down and revealed that she, too, had been sexually molested by her father for almost three years. She had been too terrified by the father's threats of reprisals to do anything about it.

The father came home one night to an empty house. Mother and daughter had moved in with friends. The niece was sent back to her own mother, still recuperating from her respiratory illness.

Rather than face a court case and the attendant traumatic ordeal, the woman decided to sue for divorce and forego criminal charges. The divorce was uncontested and the husband paid ample alimony. Not long after the divorce was completed, the woman developed the ovarian tumor. At the suggestion of her family physician, she began to see a psychiatrist. However, after her daughter developed ovarian symptoms similar to those of the mother, the woman went into deep depression and lost her will to live. It was at this point that I learned about the case.

What has given the woman great anguish is not just the knowledge of the abuse inflicted on her daughter and niece by her husband several years ago, but her own uncertainty about whether she did the right thing in not bringing criminal charges against him. Her lawyer did warn the husband at the time that if there should ever be a recurrence of abuse, there would be stern consequences. Even so, the mother suffers from the deepest feelings of guilt for not having prosecuted her husband to the fullest extent of the law. She feels guilty even though she realizes that the long-term psychological impact of a trial upon her daughter and niece could not be ignored, especially since both daughter and niece begged not to be exposed to a court ordeal.

Dr. Mitchel Covel believes that physicians should not exempt themselves from trying to attend to nonmedical causes of illnesses.

"Doctors have to prepare themselves to do all sorts of things over and above the call of duty," he said. "They have to be marital counselors, employment agents, mediators between friends. I know that medical students and young physicians steer clear of these functions but older doctors know that the extras come with the territory."

Over the course of my tenure at the medical school, I have been asked by physicians, psychiatrists, or psychologists to see patients whose will to live is flagging or who are resistant to treatment. During an average week, I may meet with twelve to fifteen such patients. Most of them are in the grip of serious illness.

Some of them have deep underlying problems. I am not a psychiatrist, nor have I had professional training as a psychologist. Whenever possible, I try to refer patients for the kind of treatment that is beyond my competence. Sometimes I try to arrange appointments for patients to see people outside the medical community, as in the case of the broker in search of a job. But I confess that, in most cases, I am badly torn. I don't want patients to think I am trying to pass them along or that I am unresponsive to their needs, even though I cannot satisfy those needs.

What, therefore, do I do? It seems to me that the least I can do is to provide a compassionate ear, not replacing but augmenting the role of the physician, psychiatrist, or psychologist. No matter how complex or poignant any given case may be, an outstretched hand can be a lifeline. People who come in search of help may not be totally fortified when they leave, but at least they can be respectfully and sympathetically received. Just from being able to speak of their problems at length, they find that the sting of pain recedes. Sometimes a degree of help is possible. When you stand at someone's side, nourishment may be provided that can carry that person forward for a week or two. During that time something genuinely helpful may turn up.

I have the utmost sympathy not just for patients but for the doctors who are called upon to treat persons who are products of tragic circumstances. The physician is usually without means to provide anything but limited help; and he himself often feels helpless when confronted by the ills of society that impose their weight on the individual.

Sir William Osler, one of the great medical figures of the twentieth century, advised medical students not to become too emotionally involved in the predicaments of seriously ill patients, especially when the physician is unable to change the circumstances that may figure in the illness. I have no doubt that some physicians can distance themselves emotionally from the tragic cases they are asked to treat. But I also know physicians who find it difficult to avoid feelings of desolation that come from their inability to halt the downward course of disease or to address themselves effectively to basic contributory causes.

As newly minted physicians, interns especially are prone not just to long hours but to the wear and tear of emotional drain. They recognize that patients are reaching out not just for scientific medical attention but for compassionate care. It is difficult to provide that kind of sympathetic understanding without becoming deeply concerned about the outcome. And if, despite all the physician's efforts and hopes, the patient continues to fade, the effect can be not merely disturbing but demoralizing. And when the downturns occur with patient after patient, it is not surprising that the physician should experience burnout or something close to it.

I confess that I find it increasingly difficult to sustain the cumulative weight of patients who suffer from catastrophic illness. My job is to give them courage and hope, to bolster their will to live, and to try to lessen the depression that so often accompanies a serious diagnosis. But the steady progression of patients in the terminal stages of illness cuts deeply into my consciousness. It is something of an uphill climb to try to demonstrate to the patient that he or she has a measure of control beyond his or her imagining.

Since many of the patients I am asked to see are suffering from advanced, high-risk disease, the result is that I am confronted with recurring defeat. If it were not for the occasional case that comes through despite overwhelming odds, I doubt that it would be possible to move on to the next patient. In order to promote confidence in patients, you have to be confident yourself. If you want to establish your credibility with the next patient, your own belief system has to recover from the most recent loss.

Fortunately, there is restorative energy in the occasional patient who does triumph—despite predictions based on the most sophisticated diagnostic technology. Perhaps the most dramatic such case in my own experience was that of a woman, thirty-four, whose breast tumor was biopsied and found to be malignant. Her physician telephoned from San Diego to talk about the case.

"This is not a situation where surgery can be considered optional," he said. "The tumor is not an ordinary one. It is like a hand grenade under a thin sheathing of skin. The breast is badly corrugated. It's a life-threatening situation. But she is resisting the surgery. The usual reasons. Mutilation. Loss of femininity, et cetera. I'd be grateful to you if you could turn her around on this. In fact, she had mentioned your name in our discussion."

The young woman—I'll refer to her as Mrs. Young—came to the office. It became apparent to me that she was a person of high intelligence and sensitivity. It also became apparent that she felt men were altogether too casual in suggesting to women that they have their breasts removed.

I told Mrs. Young that I understood her feelings but that I had to ask myself what I would do if my wife were in her situation. One thing I would most certainly do would be to get a second opinion—and possibly a third.

"I've done that," she said.

"The next thing I would do is to try to find a surgeon who is highly recognized in the field."

"I've done that, too," she said.

"In that case," I said, "I would hope that my wife would feel that we were lucky to be living at a time when medical science had discovered a way of entering her body and plucking out the offender and freeing us for a good life together."

"But wouldn't you and your wife be fearful of the surgery?"

"What happens in surgery is not just the result of the surgeon's skills but also of how you yourself think about the surgery and the confidence you yourself bring into the operating room. You have the power to program yourself for a good result."

"Program myself? How do I program myself? Why should anything I think make a difference?"

I told Mrs. Young about the studies of patients going into surgery and the difference it made if the patient had high expectations, and looked forward to being liberated from illness, rather than being governed by forebodings—and how this seemed to translate into different outcomes.

"Do you really think I have this kind of power?"

This set an obvious stage for the hand-warming exercise. Mrs. Young was a good subject and was able to increase her hand temperature by 14 degrees. Her reaction was sheer delight and surprise.

Mrs. Young returned to San Diego and notified her physician that she was prepared to go through with the surgery. A week or so later her physician phoned to say the surgery was canceled. When they took X rays just before the scheduled surgery, they discovered that the tumor had totally disappeared.

I could hardly take in what he was saying.

"Yes," he said, "I couldn't believe it myself. I'll be glad to send the X rays. The breast is soft and supple and normal in every way."

"But it was not just an ordinary tumor. That hand grenade . . ."

"It's—well, extraordinary. But it's real. We'll watch her closely, of course, but the important thing is that it has cleared up."

The only explanation for what happened is that Mrs. Young's own cancer-fighting capability had risen to the occasion. Since she had been taking no medication at the time, the only rational explanation is that the full array of immune cells—cells that produce the body's own chemotherapy and infuse it into the cancer cells—one by one—had been able to do the job.

The obvious question to be asked was whether Mrs. Young's determination and confidence had any part in the process. When I discussed the case with my colleagues, the consensus was that her new sense of control and her desire to get the most out of the surgery replaced the dread and anxiety and unblocked her cancer-fighting mechanisms. She knew she had a part to play and she was determined to do it up to the hilt.

Another dramatic case concerned a well-known actor referred to me by Dr. Avrum Bluming, the oncologist who earlier had asked me to meet with his patient, a judge (described in Chapter 2).

The actor had cancer of the larynx. He spoke in hoarse whispers. It was an especially poignant case because his voice was his prime asset. He sank into a deep depression. It seemed to me that if he was determined to combat his illness, the place to start was by recognizing the role of depression in contributing to the onset of cancer and in impairing effective treatment. Certainly it was difficult to overcome depression but it helped if one deliberately and systematically pursued the things that had given one a rewarding life in the past. There were also exercises that helped to overcome feelings of helplessness. Hand-warming techniques, for example. Also, instead of cutting oneself off from family or friends, special efforts ought to be made in that direction.

Finally, his responsibility—his part in his treatment—was to provide an environment in which Dr. Bluming could do his best. Dr. Bluming needed his full confidence. The patient also needed to have confidence in himself.

Three months later the actor returned to the office. He brought with him the good news that his cancer was in retreat and that his voice would be restored. Dr. Bluming, who makes a point of never making negative predictions, was able to get the most not just out of his medical science but of the patient's own resources.

In general, anything that restores a sense of control to a patient can be a profound aid to a physician in treating serious illness. That sense of control is more than a mere mood or attitude, and may well be a vital pathway between the brain, the endocrine system, and the immune system. The assumed possibility is that it may serve as the basis for what may well be a profound advance in the knowledge of how to confront the challenge of serious illness.

The most dramatic accounts of recoveries from catastrophic illnesses were related by physicians themselves. One particularly memorable case was described following a talk I gave at a meeting of UCLA plastic and reconstructive surgeons. During the talk, I had referred to the young woman whose scheduled mastectomy was canceled. In the open forum following the talk, a physician spoke about his own bout with cancer. The diagnosis was unequivocal; the cancer had spread to the lungs, lymph nodes, and liver.

"My oncologist friend came right out with it and said I had perhaps three to four months to live. I said nothing to my wife about it at the time but held a dialogue with myself. I decided I was not going to accept the prediction. I was going to fight it. I was going to put all my mental and physical energies into the fight. I caressed and cajoled every one of my immune cells into doing its job. For a month or two I went downhill, lost weight but not confidence. Then, little by little, the tide began to turn. I began to regain weight and strength. When I got through the sixth month, I knew the cancer didn't have a chance. How long ago was that? It was about eight years ago."

Dr. George Solomon, a pioneer in psychoneuroimmunology, told me about one of his professors at Harvard. Stricken with cancer, with lesions in head, lungs, liver, the professor continued with his classes and reassured his friends, all of whom were aware of the fatal prognosis by the specialists.

"It was thrilling," Dr. Solomon said, "to see how powerful the fighting spirit can be. For most of a year, he battled that cancer. And he won. The most important thing he had to teach us came not out of his medical lectures but out of his own experience and example. He won against all the odds—against the predictions of the specialists and against the reports based on sophisticated technology. This sort of thing doesn't happen very often, but the fact that it happens at all is the most important thing any medical student can learn."

Such statements or testimony, especially when offered by physicians themselves, helped to dispel my feeling that it was presumptuous of me to talk about remissions or recoveries that defied the predictions of specialists.

Sometimes, physicians will suggest that the favorable outcomes I cite are not as remarkable as I think they are because these cases were probably misdiagnosed in the first place. Indeed, this argument was frequently advanced following the account of my spinal illness in the *New England Journal of Medicine.* I must have received two dozen letters or more from physicians who said it was apparent to them that my illness was not ankylosing spondylitis at all but polymyalgia rheumatica, from which complete recovery is not uncommon. Dr. Hitzig became increasingly annoyed with these letters.

"What's the matter with these chaps?" he would say. "Don't they know that polymyalgia rheumatica affects the muscles and not the joints? If they had read the article, they would know that this was a problem of the joints and not the muscles."

Again, after I had a massive heart attack in 1980, some of the physicians who heard that I was back on the tennis court playing hard singles within a couple of months contended that I must have been misdiagnosed. They said that anyone who couldn't go for more than two minutes on a treadmill without a sharp drop in blood pressure and serious S-T segment depression on the cardiograph could never stand the exertion of strenuous tennis a month or two later. But when it was pointed out that the original diagnosis was made by Dr. Kenneth Shine, one of the nation's most eminent cardiologists, the skeptics had no rejoinder.

In the period following the heart attack, I would experience breathlessness or chest pain in the act of walking a short distance in an airport, or when I was late for an appointment. Yet that same day I could play hard tennis without any chest discomfort. It became obvious that being caught up in pleasant activity had a great deal to do with cardiac capacity. The hormones activated by the experience—endorphins and enkephalins especially—contributed to my physical performance. A "second wind" is another term for the ability of the body to shift into a higher gear. One thing I know is that I never experienced a "second wind" on a treadmill machine—and I wouldn't be surprised if the same is true of most other cardiac patients.

In a sense, what we are talking about here is the body's healing system—how it asserts its balances under certain circumstances. Dr. Omar Fareed had a pertinent comment along these lines.

"Medical students are not really taught about the healing system," he said. "They are taught about disease—how to diagnose and how to treat, but they are not taught how the body goes about treating itself. They will point to the immune system and let it go at that. But healing involves not just killing off disease germs or viruses but the process of reconstruction and repair."

Dr. Fareed's remark reminded me of my own efforts to find out as much as I could about the healing system when I first arrived at the university. I had obtained some general medical textbooks and looked in the index under "healing system." I found nothing, although there were separate listings for all the body's other systems—autonomic nervous system, circulatory system, digestive system, limbic system, etc. Then I turned to Merck's one-volume guide, a standard reference book in favor with medical students. Again, I turned to the index in search of "healing system" or "human healing system." Again nothing, although the other bodily systems all had separate listings. I turned to the standard medical dictionaries. Similar failure. Next I consulted the curriculum guides of various medical schools and found no references to courses dealing explicitly with the human healing system.

When I referred to this apparent gap in discussions with members of the medical faculty, I would be told that the healing system is the totality of all the body's systems and that the term "homeostasis" is used to describe the condition in which they are all in balance and fully functioning.

Even so, all the other systems are connected to the totality, yet each has its own defined functions and studies. The lack of specific attention to the healing system as such seemed to me more than a casual semantic matter. It led to the error of believing that the main locus of healing is outside, not inside, the human body. And, considering Dr. Franz Ingelfinger's statement, quoted elsewhere in this book, that 85 percent of human illnesses are self-limiting, it becomes necessary to have a clearer view than now exists about the specific processes involved in healing.

Begin with the way a cut fingernail is repaired. It is necessary for the body to deliver porcelain-type cells that can be converted at that site into a substance with adamantine qualities. The body is also called upon to manufacture silklike strands (hair); flexible and stretchable sheathing (skin); ivorylike substances (teeth); prisms that can contract or expand in order to deal with wide ranges of light intensity (eyes); solid structures for supporting weight and facilitating locomotion (bone). All these disparate substances are not delivered whole from a distant source but are manufactured on site. They are regulated from a central switchboard in the brain that is able to call upon the body's pharmacy to supply the chemicals that can be converted into the wide array of specific substances that give the human body its structural and functional characteristics.

An equally mysterious but important function is the way the body attends to its own growth. It is well known, for example, that hormones, obeying genetic instructions, can cause bodily structures to expand—longer legs and arms, wider shoulders, bigger hands and feet, etc. But, while this process can be observed and measured, the mechanisms are only vaguely understood.

How important is such understanding? Shouldn't we be content to recognize the existence of a process without having to penetrate all its secrets? The answer is that expanding knowledge and human progress go together. True, the human species has survived until now with incomplete knowledge of growth and healing. But we may do a much better job once we understand the "why" of things. We may not know exactly where new knowledge will lead, only that we will get nowhere of consequence without it. "Whatever you cannot understand," said Goethe, "you cannot possess." The possession by human beings of self-knowledge opens the door to knowledge of the universe. "The proper study of mankind," said Pope, "is man."

The highest exercise of a physician's skills is to prescribe not just out of a little black bag but out of his or her knowledge of the human healing system. The ability of the physician in unblocking or enhancing or augmenting this healing system will constitute the grand confluence of the art and science of medicine.

10

THE LAUGHTER CONNECTION

"The most acutely suffering animal on earth invented laughter."

—FRIEDRICH NIETZSCHE

"There ain't much fun in medicine, but there's a heck of a lot of medicine in fun."

—JOSH BILLINGS

The telephone call was from the Associated Press correspondent stationed in Chicago.

"How does it feel to be fully vindicated?" he was asking.

I drew a blank and said so.

"There's a report in the new issue of *JAMA* [*Journal of the American Medical Association*] providing scientific evidence that you were right when you wrote that laughter is useful in combating serious illness," he continued. "It's by Swedish medical researchers whose experiments show that laughter helps the body to provide its own medications. Let me read from the report: 'A humor therapy program can improve the quality of life for patients with chronic problems. Laughter has an immediate symptom-relieving effect for these patients.'

"Since you were heavily criticized by some physicians when your article first appeared in the *New England Journal of Medicine*, it must feel pretty good to have this verification."

Of course I was gratified by the report in *JAMA*. In *Anatomy of an Illness*, first published in 1976, I had reported my discovery that ten minutes of solid belly laughter would give me two hours of pain-free sleep. Since my illness involved severe inflammation of the spine and joints, making it painful even to turn over in bed, the practical value of laughter became a significant feature of treatment.

Dr. William Hitzig, my physician, was as fascinated as I was by the clear evidence that laughter could be a potent painkiller. He tested this proposition by comparing my sedimentation rate before and after my response to amusing situations in films or books. The sedimentation test measures the extent of inflammation or infection in the body. Since my sedimentation rate was in the upper range, any reduction was to be welcomed. Dr. Hitzig reported to me that just a few moments of robust laughter had knocked a significant number of units off the sedimentation rate. What to him was most interesting of all was that the reduction held and was cumulative.

Even more encouraging, the retreat of pain was accompanied by a corresponding increase in mobility. At that time, little was known about the ability of the human brain to produce secretions with morphinelike molecules—endorphins and enkephalins. Looking back now, in the light of this knowledge, I realize that the laughter probably played a part in activating the release of endorphins.

In writing about this experience, I was careful to point out that I didn't regard the use of laughter as a substitute for traditional medical care. I also emphasized that I tried to bring the full range of positive emotions into play—love, hope, faith, will to live, festivity, purpose, determination.

Obviously, what worked for me may not work for everyone else. Accumulating research points to a connection between laughter and immune enhancement, but it would be an error and indeed irresponsible to suggest that laughter—or the positive emotions in general—have universal or automatic validity, whatever the circumstances. People respond differently to the same things. One man's humor is another man's ho-hum. The treatment of illness has to be carefully tailored to suit the individual patient.

But not unnaturally, perhaps, the laughter aspects of the recovery made good newspaper copy. I was disturbed by the impression these accounts created that I thought laughter was a substitute for authentic medical care. In fact, my principal reason for writing the article about my illness in the *New England Journal of Medicine* was to correct this impression. I emphasized that my physician was fully involved in the process and that we regarded laughter as a metaphor for the full range of the positive emotions.

Perhaps I might have been a lot less defensive if I had known then what I know now. Medical researchers at a dozen or more medical centers have been probing the effects of laughter on the human body and have detailed a wide array of beneficial changes—all the way from enhanced respiration to increases in the number of disease-fighting immune cells. Extensive experiments have been conducted, working with a significant number of human beings, showing that laughter contributes to good health. Scientific evidence is accumulating to support the biblical axiom that "a merry heart doeth good like a medicine."

Of all the gifts bestowed by nature on human beings, hearty laughter must be close to the top. The response to incongruities is one of the highest manifestations of the cerebral process. We smile broadly or even break out into open laughter when we come across Eugene Field's remark about a friend "who was so mean he wouldn't let his son have more than one measle at a time." Or Leo Rosten's reply to a question asking whether he trusted a certain person: "I'd rather trust a rabbit to deliver a head of lettuce." Or, as Rosten also said, "Let's go somewhere where I can be alone." Or Evan Esar's definition of love: "A comedy of Eros." These examples of word play illustrate the ability of the human mind to jump across gaps in logic and find delight in the process.

Surprise is certainly a major ingredient of humor. Babies will laugh at sudden movements or changes in expression, indicating that breaks in the sequences of behavior can tickle the risibilities. During the days of the silent films, Hollywood built an empire out of the surprise antics of its voiceless comedians—Harold Lloyd swinging from the hands of a giant clock, Charlie

Chaplin caught up in the iron bowels of an assembly belt, or Buster Keaton chasing a zebra.

It has always seemed to me that laughter is the human mind's way of dealing with the incongruous. Our train of thought will be running in one direction and then is derailed suddenly by running into absurdity. The sudden wreckage of logical flow demands release. Hence the physical reaction known as laughter.

Consider the story of the two octogenarians on a park bench. One asks the other:

"Do you believe in reincarnation?"

"Well, Joe," replies Harry, "I've never really thought much about it."

"Maybe we ought to start thinking about it," says Joe. "One of us is going to go first. Let's agree that the one who is left behind will come to this park bench every Wednesday at eleven A.M. and the one who has departed will find a way of getting a message to him at that time about reincarnation and all those other things that are beyond our ken."

Harry agrees.

One month later, Joe dies peacefully in his sleep. Every week for several months Harry takes up his station at the park bench at 11:00 A.M.

Then one Wednesday at the appointed hour, he hears a voice, as from afar.

"Harry, Harry, can you hear me?" the voice says. "It's Joe."

"Joe, for heaven's sake, what is it like?"

"You wouldn't believe it, Harry, about the only thing you do up here is make love. They wake you up at seven in the morning and you make love until noon. After lunch and a nap you're at it again right through until dinnertime."

"Good Lord, Joe, what are you and where are you?"

"I'm a rabbit in Montana!"

The brain accommodates itself to the collision of logic and absurdity by finding an outlet in the physiological response we recognize as laughter.

Does humor make a difference in one's approach to challenging situations? According to some medical studies, public speaking is one of the most stressful experiences in any catalogue of human activity. I have no way of confirming that fact scientifically, but I can attest, on the basis of some two thousand lectures during the past forty years, that both speaker and audience seem to get along much better with a little humor.

The first formal lecture I gave was in New York City's Town Hall, then the mecca for public speakers. After I completed the talk, Bennett Cerf, the book publisher and collector of jokes, came up to greet me.

"That was a good talk," he said, "considering it's your first try. But you're a little too serious. You've got to warm up your audience. Start with a humorous story. It's always a good idea to involve yourself in the anecdote, especially if a celebrity or two figures in the story. Hasn't anything funny happened to you recently that also involves someone who's famous?"

I told Bennett I had spoken in Albany only a week earlier at a special convocation of the Board of Regents to welcome General Eisenhower in his new role as president of Columbia University. I sat next to General Eisenhower and tried to compose myself for my talk. My discomfiture must have been apparent to the general. He leaned toward me and spoke in a whisper.

"What's the matter? You look a little pale."

I whispered back that my appearance was not deceptive, and that the prospect of speaking to so many educators in their university gowns was a little intimidating.

"Do as I do," he said. "Whenever I feel nervous before I speak I use a little trick."

"Trick? What trick?"

"I just transfer my nervousness to the audience," he said.

"And how do you do that?"

"Very simple. I look out at all the people in the audience and just imagine that everyone out there is sitting in his tattered old underwear!"

Bennett Cerf was entranced with this account.

"Great, great!" he exclaimed. "Now, the next time you speak, just begin your talk with the Albany story and the advice given you by General Eisenhower. You'll see what a difference it makes."

Two weeks later I spoke in St. Louis and began my talk with the Eisenhower anecdote. It didn't produce even a ripple. The audience was stony faced. The rest of my talk was uphill all the way.

Following the lecture, a man came up to me.

"That story you told about General Eisenhower," he said. "Are you sure it happened to you?"

"Yes," I said. "Of course it did."

"That's strange," he replied. "Bennett Cerf lectured here last week and said it happened to him."

Even so, I can testify that Bennett Cerf gave me good advice. People who come out to hear a talk tend to be a little more attentive if the speaker illustrates his points with occasional amusing references or anecdotes. If, on the other hand, you keep pounding away at an audience with solemn pronouncements, you can expect to be rewarded with a collective glaze.

Cerf was right, too, about the strategy of involving yourself in the stories. I got a lot of mileage at the beginning of my talks by referring to something that happened to me when I went to Kansas City on a rainy night for a talk at the Convention Hall. The structure consisted of several auditoriums of various sizes—a prizefight arena and a hall capable of accommodating six or seven thousand people.

I went in through the main entrance, checked my raincoat, and went to the third-floor auditorium, as specified on my itinerary sheet. I walked down the side aisle and noticed that the stage had no approaching steps. I asked a member of the audience if he could tell me how to get backstage. He pointed to a side door, then told me to go inside and turn left.

As directed, I went through the designated door—then spun around to catch it behind me, thinking it might be bolted from the inside. I was too late. The knob didn't turn. In the dim light I could see that there was no passageway to the left—something I should have realized about Kansas City in the first place—only a stairway, the bottom of which was lost in darkness.

I rapped on the door behind me; it had a heavy metal thickness and I knew that continued rapping would produce only sore knuckles.

There was nothing to do except to make the great descent. I came to the bottom, turned left into the darkness, and found myself stumbling through the remains of old stage sets. Eventually, I spied a thin line of light along the floor and groped my way toward it, thinking it might be a door. I was right. I turned the knob, opened the door, and stepped out into the brilliantly lighted platform of a poultry convention being addressed by H. V. Kaltenborn. A gasp went up from the audience. Mr. Kaltenborn turned around, recognized me, and asked what was happening.

"I'll explain later," I said, and backed out again into the catacombs of Kansas City, resuming my encounters with the old stage sets. Eventually, I came to another door. It had a bar handle. I pushed, the door opened, and I found myself on the street on the far side entrance to the building.

The rain had become a downpour. I had to choose between a retreat back into the catacombs or a wet platform presence. It was an easy choice. I made a run for it. Several hundred yards later, I reentered the structure and proceeded to the third-floor auditorium, where an attendant asked for my ticket.

"I realize I don't look much like it at this point, or perhaps at any point," I said, "but I happen to be the speaker."

The attendant looked at the thoroughly soaked figure in front of him.

"Look, bud," he said, "the speaker went inside twenty minutes ago."

His facts were unassailable.

"How do I get in?" I asked wearily.

The attendant pointed to the box office.

"You pay your way in, same as everyone else," he said.

I did as directed but not without resentment, being forced to pay three dollars for a talk I had heard before. I perceived, however, a certain redeeming feature of this episode. It occurred to me that the country might have far better lectures if the lecturers occasionally had to pay their way in to hear themselves.

Such misadventures give me ammunition for the beginning of my talks. Audiences seem to relish accounts involving mishaps to speakers. The Kansas City story puts them in a relaxed and receptive mood.

Dr. Novera Herbert Spector, the neurophysiologist, sent me an account written more than a half-century ago of the physiological benefits of laughter. It was by Dr. James Walsh, then medical director of the School of Sociology at Fordham University in New York.

Dr. Walsh reported on his research showing that hearty laughter stimulates internal organs, "making them work better through the increase of circulation that follows the vibrating massage that accompanies it, and heightens resistive vitality against disease." Dr. Walsh then particularized by detailing the beneficial effects of laughter on lungs, liver, heart, pancreas, spleen, stomach, intestines, and the brain. He wrote that laughter has the effect of brushing aside many of the worries and fears that set a stage for sickness.

Dr. William Fry, Jr., of the Department of Psychiatry at Stanford Medical School, likens laughter to a form of physical exercise. It causes huffing and puffing, speeds up the heart rate, raises blood pressure, accelerates breathing, increases oxygen consumption, gives the muscles of the face and stomach a workout, and relaxes muscles not involved in laughing. Twenty seconds of laughter, he has contended, can double the heart rate for three to five minutes. That is the equivalent of three minutes of strenuous rowing.

Just in psychological terms, laughter can confer benefits. In the 1950s, Dr. Gordon Allport, psychologist-teacher at Harvard University, theorized that humor, like religion or climbing a mountain, provides a new perspective on one's place in society. Dr. Annette Goodheart, a psychologist in Santa Barbara, confirmed this view in her own work. She found that humor helps an individual to confront personal problems in a more relaxed and creative state. Along the same line, Dr. Alice M. Isen, of the Department of Psychology and Johnson Graduate School of Management at Cornell University, said that laughter increases creativity and flexibility of thought. One of her studies involved individuals who failed to tack a candle onto a corkboard wall in such a way as to not drip wax on the floor while it burned. Those persons who had just seen a short comedy film were better able to devise innovative approaches to the task than those who had not. She felt that the group went from "functional fixedness" to "creative flexibility" after seeing the film. In tests of mental acuity, Isen has also found that positive feelings, induced by humorous films or the giving of a candy gift, enable persons to use language more colorfully.

Laughter has also been shown to be helpful in reducing "discomfort sensitivity." Drs. Rosemary and Dennis Cogan, et al., of the Department of Psychology at Texas Tech University in Lubbock, randomly assigned forty undergraduates into four groups: (1) laughter, (2) relaxation, (3) informative narrative, and (4) no-treatment control. Members of groups 1, 2, and 3 listened individually to a twenty-minute audio tape appropriate to their group. Discomfort was produced by pressure from the automatic inflation of a blood pressure cuff. Subjects could withstand the highest level of discomfort after being exposed to humorous material; the next highest after a relaxation technique. The Cogans and their colleagues suggest that humor can be particularly useful in alleviating the pain of injections a patient must receive, as well as postoperative pain.

If laughter can improve one's perspective on life and on pain, it follows that it might be helpful in combating unusual stress. Drs. Rod A. Martin and Herbert M. Lefcourt, then of the University of Waterloo in Ontario, Canada, studied the connection between humor and adjustment to major life stresses. They gave fifty-six undergraduates four tests designed to measure the capacity to enjoy humor under a variety of circumstances. Three out of four tests showed that those who valued humor the most were also most capable of coping with tensions and severe personal problems.

Drs. Martin and Lefcourt subsequently studied sixty-seven students and found that those who had the greatest ability to produce humor "on demand" in impromptu routines are also best able to counteract the negative emotional effects of stress.

Dr. James R. Averill, then of the Department of Psychology, University of California, Berkeley, measured a number of physiological reactions accompanying various emotional states. Fifty-four undergraduates were divided into three groups—a sadness group, a mirth group, and a control group. The sadness group watched a film on John F. Kennedy; the mirth group saw a Mack Sennett–type silent comedy; and the control group viewed a non-arousing documentary film. Measurements of blood pressure, heart rate, finger and face temperature, finger pulse volume, skin resistance, and respiration were taken during the showing of the films. Averill found that laughter produced respiratory changes, and that sadness yielded changes in blood pressure. The main conclusion of the study was that emotions produce measurable physiological change.

Dr. Paul Eckman and colleagues, of the University of California at San Francisco, School of Medicine, measured physiological differences during six emotional states—surprise, disgust, sadness, anger, fear, and happiness. They asked sixteen individuals—actors and scientists—to mimic prototypical emotional facial expressions and then to experience each of the six emotions by reliving a past emotional experience while measurements of heart rate, hand temperature, skin resistance, and muscle tension were taken.

The results were striking. Not only did significant physiological differences between the negative emotions and the positive emotions turn up, but different emotions produced different effects. For example, anger was characterized by a *high* heart rate and hand temperature increases, whereas fear was accompanied by high heart rate increases and *decreases* in hand temperature. Happiness was associated with *low rises* in heart rate and hand temperature.

Some of the researchers involved in these studies sent me reports of their work, saying they had been prompted by my experience as reported in *Anatomy of an Illness* to seek scientific evidence for the value of laughter. Equally gratifying were the accounts from hospitals of new facilities that featured humor and creativity as integral parts of the hospital program.

The first to respond was St. Joseph's Hospital in Houston, Texas. I received a telephone call from Dr. John Stehlin, oncological surgeon and medical researcher, asking if I would come to Houston to participate in the dedication ceremonies of a new feature of the hospital called the "Living Room." He said that the cancer floor had been redesigned to accommodate a large room furnished with easy chairs, hi-fi equipment, an art corner, video and audio sets, and a library.

"You can't imagine a setting more unlike a hospital," he said. "You would enjoy seeing how a pleasant environment can brighten the mood of the patients. Amusing films are one of our main props. You will enjoy meeting the nuns. They like the idea of making laughter a regular part of the hospital's philosophy."

Three weeks later, I went to Houston for the dedication ceremonies. The entire hospital was in a festive mood for the occasion. The media exhibited a strong interest in the Living Room and the idea behind it. TV crews were setting up their cameras at the entrance to the hospital and in the Living Room itself.

I judged Dr. Stehlin to be in his mid-fifties. His appearance was tall, trim, athletic, hearty. He took me into his office and told me that the hospital's experiment with laughter had been highly successful. The patients were relaxed and responsive; they appeared to be making fewer demands on the hospital staff. I met with at least a dozen patients whose arms or legs had been saved from amputation by procedures developed by Dr. Stehlin and his associates. The affected limb would be tourniqueted, protecting the rest of the body from the chemotherapy, which was then free to attack the malignancy without penalty to the rest of the body.

The Living Room was in high favor with both ambulatory and wheelchair patients. They kept the videotape machine busy, viewing comedy films that Dr. Stehlin was able to obtain directly from the film companies with their compliments.

Dr. Stehlin said that he found himself stealing time away from his desk to look at the comedies.

"You have no idea how good it makes me feel just to see the patients laughing and enjoying themselves," he said. "This place is getting to be unrecognizable as a hospital—and that's all to the good. Now, before we go to the Living Room, I want you to meet the nuns, especially Mother Romano. She's almost eighty but is spry and energetic. She has fallen in with this new approach and makes a point of telling the patients funny stories. Just yesterday she was telling them about an incident at a nightclub. It seems that the headwaiter was looking around the room and saw a man slide off his chair. He rushed over to the table.

" 'Madam,' he said, 'your husband has slipped off his chair and fallen under the table.'

" 'Sir,' she said, 'my husband has just come into the room.' "

I trailed after Dr. Stehlin as he led me to the Living Room, stopping along the way at the rooms of cancer patients, where he introduced me, saying that *Anatomy of an Illness* was the inspiration for the Living Room.

We met Mother Romano, a wisp of a woman but a human dynamo. She pumped my arm vigorously when we were introduced and rhapsodized over the patients' response to the Living Room. She accompanied us on visits to other cancer patients, then looked at her watch and said the dedication ceremonies were about to begin.

The Living Room was jammed with people—patients, hospital personnel, trustees, civic leaders, press representatives. I guessed there were between 150 and 200 people. The room was brightly decorated. As Dr. Stehlin had said, it was the antithesis of everything that is popularly associated with a hospital.

Dr. Stehlin spoke first, then introduced Mother Romano, who described the auspicious effect of the new room not only on the patients but on the nurses and other members of the hospital staff. Then Dr. Stehlin introduced me to the audience.

I said the obvious things—how grateful I was that my hospital experience played a part in the decision to create the new facility at St. Joseph's. I described the research going forward at UCLA and elsewhere showing that laughter could produce auspicious physiological change.

The meeting with Dr. Stehlin led to a close friendship. We were able to use some of the funds placed at my disposal to support his research at the hospital in studying the salutory effects of the psychological environment surrounding the treatment of cancer patients.

The Living Room at St. Joseph's was only the first of two dozen or more similar programs at hospitals throughout the country. At St. John's hospital in Los Angeles, a special channel on the TV sets in each room enabled the patients to turn on comedy films day or night. As at St. Joseph's, the hospital acquired a considerable library of amusing motion pictures. I relished the opportunity given me by the nuns at St. John's to dedicate the new comedy channel.

I participated in six or seven similar ceremonies at hospitals in the Los Angeles area. At the L.A. County Hospital, Joseph Barbera, creator of Huckleberry Hound and Yogi Bear, presented the children's division with life-size replications of the characters made famous in his TV cartoons.

UCLA redecorated the hospital's children's floors, replacing the drab whiteness with bright colors and amusing drawings. UCLA also used art throughout the entire hospital in a program devised by Devra Breslow.

Cedars Sinai Hospital, in Los Angeles, accepted the proposal of Marcia Weismann, a prominent art collector, to decorate the walls of the hospital with superb reproductions of outstanding art. Mrs. Weismann also arranged for art lectures at the hospital.

Good Samaritan Hospital of Los Angeles instituted a humor channel for patients' TV sets, along the lines of the program instituted at St. John's. Dr. David Cannom, head of cardiology at Good Samaritan, arranged for piano concerts at the hospital by Mona Golabek, an outstanding local artist.

Perhaps the most far-reaching program of all in the use of not just humor but also music, art, and literature in the treatment of the seriously ill was instituted by the Duke University Comprehensive Cancer Center in Durham, North Carolina. The word "Comprehensive" is to be taken literally. All the aspects of treatment are taken into account—the emotional needs of the patient; the interests or hobbies of the patient that can improve the climate of medical care, inside or outside the hospital; the use of a "laugh wagon" that is hardly less in evidence in the corridors of the hospital than the "pill carts" of the nurses; the services of volunteers to provide companionship or solace or to carry out errands for the patients—all these are integral to Duke's program. Dr. Robert Bast, director of the center, believes that the psychosocial factors are no less important than medical prescriptions in the structure of a hospital and medical school.

At Duke, I learned of two specific examples of the way humor can be used to lessen problems caused by illness.

A patient at the university hospital suffered from a severe case of otitis media (inflammation of the middle ear). The nurse, recognizing that the emotional state of the patient was important, read to him out of a collection of stories by Woody Allen. The patient's sense of delight blocked any nervousness, which the nurse felt might have interfered with the efficacy of the medication. Whether or not her strategy could be verified by physiological measurements, her awareness that psychological factors could enhance or impair medical treatment won the commendation of the hospital authorities.

Another example concerned a man who suddenly developed what is known as a paroxysmal tachycardia—a wildly beating heart. Ordinarily, the condition is temporary but when it is also marked by an irregular pulse, prompt medical attention is required.

The man's wife telephoned the family physician, who said he would rush over to the house. In the meantime, he instructed the wife to do everything possible to keep the patient free of panic, the effects of which could intensify and complicate the underlying problem. The woman played a videotape of "Candid Camera" reruns, one sequence of which showed Buster Keaton at a lunchroom counter struggling to keep his eyeglasses and his toupee from falling into his soup. Her husband fell to laughing over the absurdity and completely blotted out any trace of panic. By the time the doctor arrived, the pulse had settled back to normal. The doctor credited the wife with having helped to avert a potentially serious situation.

I must not overlook the "Lively Room" at the DeKalb Hospital in Decatur, Georgia, or the new floor of the Shawnee Mission Hospital near Kansas City, Missouri, both of which pay careful attention to the atmosphere of the hospital in assessing the effectiveness of patient treatment. Humor and creativity are important parts of the program.

A number of hospitals are subscribing to a nonprofit service established by Bea Ammidown Miller of Los Angeles, who had survived a devastating automobile accident. Mrs. Miller devised a cart containing all sorts of humorous materials that can be wheeled directly into the rooms of patients—a sort of laughter smorgasbord. The project is called Humor/x and has caught on in a number of hospitals across the country.

Positive emotions may have beneficial effects during exercise. Dr. Gary E. Schwartz, of the Yale University Department of Psychology, asked thirty-two college volunteers who had had some acting experience to imagine events that would evoke happiness, sadness, anger, fear, relaxation, and neutral emotion, and then had the participants sustain those emotions while exercising. Each emotion produced different blood pressure, heart rate, and body movement patterns during imagery alone and with exercise. The researchers found that anger during exercise raised the participants' heart rates an average of thirty-three beats per minute—more than double the increase during neutral (or normal) exercise; whereas a context of relaxation raised participant heart rates less than half the average increase of normal exercise. In addition, sadness was the only condition found to suppress the normally expected increases in heart rate and blood pressure activity during exercise.

Other researchers have added to the research evidence on the salutary effects of laughter. Dr. Kathleen M. Dillon, of Western New England College, measured salivary immunoglobulin A (sIgA) concentrations in ten student volunteers before and after they viewed humorous and nonhumorous thirty-minute videotapes. Salivary immunoglobulin A is believed to have a protective capacity against some viruses. Dr. Dillon discovered that these protective concentrations increased significantly after the students viewed the humorous videotape. The concentrations remained unchanged after the nonhumorous tape. She also found that the individuals who said they turned to humor as a way of coping with difficult life situations also had the highest initial levels of the protective concentrations. Her conclusion was that a consistently cheerful approach to life bolstered the body's disease-fighting forces. She replicated this finding in a study of nursing mothers.

Dr. David C. McClelland, formerly of Harvard University and later of Boston University, found that students who viewed a fifty-minute film of Mother Teresa (designed to induce a positive, caring emotional state) experienced significant increases in sIgA concentrations, whereas those who viewed a powerful documentary film about the Nazis in World War II (designed to elicit feelings of anger in the viewer) did not show appreciable changes in sIgA. In a subset of Mother Teresa viewers, the gains in sIgA were sustained beyond one hour if mental exercises dwelling on loving and being loved were conducted immediately after the film was shown.

I shared some of the research funds available to me with Dr. Lee S. Berk of the Loma Linda University Medical Center in California, to help support his studies of the way laughter might enhance the immune system. He and his colleagues measured changes in several "stress" hormones in ten healthy male subjects after they viewed a sixty-minute humorous film. Dr. Berk hypothesized that laughter has beneficial effects on the immune system. Indeed, in a subsequent study, he found a significant increase in spontaneous blastogenesis (immune cell proliferation) accompanied by a marked decrease in cortisol, a hormone that has an immune-suppressing capability.

Articles have appeared in the national press reporting scientific verification of the usefulness of laughter in combating illness and promoting health. Jane Brody, the highly regarded writer on health matters, used her column in *The New York Times* to report on the "growing number of physicians, nurses, psychologists, and patients who have used the uniquely human expression of mirth to reduce stress, ease pain, foster recovery, and generally brighten one's outlook on life, regardless of how grim the reality."

Jane Brody also called attention to the new practices of nursing homes and hospitals in using "laughter wagons" stocked with humorous materials and "other gimmicks likely to amuse patients." She reported that members of a group at Oregon Health Sciences University calling themselves "Nurses for Laughter," wear buttons that read: "Warning: Humor may be hazardous to your illness."

Jane Brody quoted Dr. Marvin E. Herring of New Jersey's School of Osteopathic Medicine as saying that "the diaphragm, thorax, abdomen, heart, lungs, and even the liver are given a massage during a hearty laugh." The expression used in *Anatomy of an Illness* was "internal jogging."

Jane Brody made some practical suggestions:

> Instead of flowers, consider sending the patient a funny novel, a book of jokes, a silly toy, a humorous audio tape and portable recorder or, if a video recorder is available, a funny movie. When my best friend contracted a life-threatening disease, I made her a looseleaf "book of laughs" stuffed with *New Yorker* cartoons, classic witticisms, and personalized homemade jokes. Years after recovering, she continues to use the joke book whenever she thinks she is getting sick.
>
> Brighten the sick room with mobiles, homemade silly sculptures, comical photos, and get-well cards. Place a poster of a scenic view on the window or wall and change it often.

Keep on the lookout for humorous happenings and statements you can tell the patient about. Arrive at the bedside with a funny story instead of a complaint about the terrible traffic or parking problem.

Seek out caretakers with a sense of humor. . . .

Consider organizing a local scout troop, school, or senior citizens group to prepare riddles or jokes that can be placed on patients' breakfast trays. Or challenge patients to come up with their own humorous captions for certain drawings. . . .

Finally, when you hear a good joke, write it down or quickly relate it to someone to help you remember it.

In a similar vein, the *News Bulletin* of the American Association of Retired Persons, reported that Kaye Ann Herth, a nurse in Tulsa, Oklahoma, takes a "funny-bone history" of patients in an effort to ascertain their taste in humor. She uses this information to write "prescriptions" for laughter tailored to patient taste.

The *News Bulletin* also reported that volunteers for the Andrus Gerontology Center at the University of Southern California produced a handbook on the use of humor in long-term care facilities. The handbook gave these examples of humorous one-liners:

"A man's home is now his hassle."

"It's better to have loved a short girl than never to have loved a tall."

"Dieting—the triumph of mind over platter."

These one-liners, the article said, may or may not be funny but they did have a good effect on the patients, many of whom were "coaxed out of their shells" and took an active part in the programs.

Earlier in this chapter, I spoke of the role of humor in creating a relaxed and responsive environment in audiences. I had an opportunity to see this effect in heightened form at the Sepulveda (California) Veterans' Administration Hospital.

I had gone to the hospital following a meeting with physicians assigned to the cancer unit. They were concerned because the mood of the cancer patients was so bleak that they feared the collective environment of treatment was being impaired.

At the suggestion of the doctors I met with the veterans in the cancer unit. There were perhaps fifty or sixty of them. They sat in rows and were every bit as glum as I had anticipated.

I reported on my conversation with their doctors, and said I doubted that they were helping them or themselves with the grim mood of the place. Certainly one could understand the reason for their feelings—and it was arrogant for anyone to lecture to them about it. But in coming to Sepulveda they were reaching out for help—and they were entitled to know what would optimize that prospect.

Any battle with serious illness, I said, involved two elements. One was represented by the ability of the physicians to make available to patients the best that medical science has to offer. The other element was represented by the ability of patients to summon all their physical and spiritual resources in fighting illness.

I said I hoped the veterans would agree that their part of the job was to create an environment in which the doctors could do their best. One thing they might do to replace the grim atmosphere was to put on performances. We could give them scripts of amusing one-act plays. Some of them might wish to produce or direct or act. If they wished, we could help them obtain videocassettes of amusing motion picture films. Ditto, audio cassettes of stand-up comics. One way or another, their part in the joint enterprise with their doctors was to create a mood conducive to the best medical treatment obtainable.

The veterans accepted the challenge. When I returned to the hospital several weeks later and spoke to the doctors I was pleased to have them describe the change not just in the general environment but in the mood of the individual patients.

When I met with the veterans, they no longer sat in rows. They sat in a large circle. They were part of a unity; they could all see one another. When they began their meeting, each veteran was obligated to tell something good that had happened to him since the previous meeting.

The first veteran spoke of his success in reaching by telephone a buddy he had not seen since the Korean War. He had tracked his buddy to Chicago and finally made the connection. They spoke for a half-hour or more. And the good news was that his buddy was coming to visit him in California.

Cheers.

The next veteran read from a letter he had received from a nephew who had just been admitted to medical school. He quoted the final sentence of the letter:

"And, Uncle Ben, I want you to know that I'm going into cancer research, and I'm going to come up with the answer, so you and your buddies just hang in there until I do."

More cheers.

And so it went, each person at the meeting taking his turn. Then I discovered that everyone was looking at me and that I was expected to report on what it was that was good that had happened to me.

I searched my recent memory and realized that something quite good had in fact happened to me only a few days earlier.

"What I have to report is better than good," I said. "It's wonderful. Actually, it's better than wonderful. It's unbelievable. And as long as I live, I don't expect that anything as magnificent as this can possibly happen to me again."

The veterans sat forward in their seats.

"What happened is that when I arrived at the Los Angeles airport last Wednesday my bag was the first off the carousel."

An eruption of applause and acclaim greeted this announcement.

"I had never even met anyone whose bag was the first off the carousel," I continued.

Again, loud expressions of delight.

"Flushed with success, I went to the nearest telephone to report my arrival to my office. That was when I lost my coin. I pondered this melancholy event for a moment or two, then decided to report it to the operator.

" 'Operator,' I said, 'I put in a quarter and didn't get my number. The machine collected my coin.'

" 'Sir,' she said, 'if you give me your name and address, we'll mail the coin to you.'

"I was appalled.

" 'Operator,' I said, 'I think I can understand the reason

behind the difficulties of A.T.&T. You're going to take the time and trouble to write down my name on a card and then you are probably going to give it to the person in charge of such matters. He will go to the cash register, punch it open and take out a quarter, at the same time recording the reason for the cash withdrawal. Then he will take a cardboard with a recessed slot to hold the coin so it won't flop around in the envelope. Then he, or someone else, will fit the cardboard with the coin into an envelope, first taking the time to write out my address on the envelope. Then the envelope will be sealed. Someone will then affix a twenty-cent stamp on the envelope. All that time and expense just to return a quarter. Now, operator, why don't you just return my coin and let's be friends.'

" 'Sir,' she repeated in a flat voice, 'if you give me your name and address, we will mail you the refund.'

"Then, almost by way of afterthought, she said, 'Sir, did you remember to press the coin return plunger?'

"Truth to tell, I had overlooked this nicety. I pressed the plunger. To my great surprise, it worked. It was apparent that the machine had been badly constipated and I happened to have the plunger. All at once, the vitals of the machine opened up and proceeded to spew out coins of almost every denomination. The profusion was so great that I had to use my empty hand to contain the overflow.

"While all this was happening, the noise was registering in the telephone and was not lost on the operator.

" 'Sir,' she said, 'what is happening?'

"I reported that the machine had just given up all its earnings for the past few months, at least. At a rough estimate, I said there must be close to four dollars in quarters, dimes, and nickels that had just erupted from the box.

" 'Sir,' she said, 'will you please put the coins back in the box.'

" 'Operator,' I said, 'if you give me your name and address I will be glad to mail you the coins.' "

The veterans exploded with cheers. David triumphs over Goliath. At the bottom of the ninth inning, with the home team behind by three runs, the weakest hitter in the lineup hits the ball out of the park. A mammoth business corporation is brought to its knees. Every person who had been exasperated by the loss of a coin in a public telephone booth could identify with my experience and share both in the triumph of justice and the humiliation of the mammoth and impersonal oppressor.

The veterans not only were having a good time; they were showing it in their relaxed expressions and in the way they moved.

One of the doctors stood up.

"Tell me," he said, "how many of you, when you came into this room a half hour or so ago, were experiencing, more or less, your normal chronic pains?"

More than half the veterans in the room raised their hands.

"Now," said the doctor, "how many of you, in the past five or ten minutes, discovered that these chronic pains receded or disappeared?"

The same hands, it appeared to me, went up again.

Why should simple laughter have produced this effect? Brain researchers with whom I have spoken have speculated that the laughter activated the release of endorphins, the body's own pain-reducing substance. The veterans were experiencing the same effects that had occurred to me in my own bout with inflammatory joints many years earlier. The body's own morphine was at work.

In view of what is now known about the role of endorphins not only as a painkiller but as a stimulant to the immune system, the biological value of laughter takes on scientific validity.

"If you wish to glimpse inside a human soul and get to know a man," Dostoevski writes in his novel *The Adolescent*, "don't bother analyzing his ways of being silent, of talking, of weeping, or seeing how much he is moved by noble ideas; you'll get better results if you just watch him laugh. If he laughs well, he's a good man. . . .

"I consider it one of the most important conclusions derived from my life experience," Dostoevski continues. "I especially recommend it to the attention of young would-be brides who are prepared to marry the man of their choice but are still watching him with misgivings and distrust and cannot take the decisive step. All I claim to know is that laughter is the most reliable gauge of human nature. Look at children, for instance: Children are the only human creatures to produce perfect laughter and that's just what makes them so enchanting. I find a crying child repulsive whereas a laughing and gay child is a sunbeam from paradise to me, a revelation of future bliss when man will finally become as pure and simple-hearted as a babe."

One of the most frequent questions put to me by patients: "Where do you find things to laugh at?"

My favorite sources are books—the very substantial humor collections by Isaac Asimov and Leo Rosten and, especially, E. B. White's *Subtreasury of American Humor*. (Appended to this chapter is an excellent list of books, audio-cassettes, and videocassettes compiled by the Comprehensive Cancer Center of Duke University.)

While I was a hospital patient, I had a good time viewing old motion pictures and TV programs. I found some of the old Marx Brothers comedies and "Candid Camera" reruns highly useful for this purpose. But I also had access to other materials, including classified notices from the "Personals" columns of the *Saturday Review*.

It is a serious error to suppose that laughter is the only emotional antidote to stress or illness. Some people tend to be humor-resistant and derive benefits in other ways. An appreciation of life can be a prime tonic for mind and body. Being able to respond to the majesty of the way nature fashions its art—the mysterious designs in the barks of trees, suggesting cave paintings or verdant meadows interrupted by silvery streams; the rich and luminous coloring of carp fish with blues and yellows and crimsons seemingly lit up from within; the bird of paradise flower, an explosion of colors ascending to a triumphant and jaunty crest of orange and purple; the skin of an apple, so thin it defies measurement but supremely protective of its precious substance; the way the climbing trunk of a tree will steer its growth around solid objects coming between itself and the sun; the curling white foam of an ocean wave advancing on the shore, and the way the sand repairs and smooths itself by the receding water; the purring of a kitten perched on your shoulder, or the head of a dog snuggling under your hand; the measured power of Beethoven's *Emperor* Concerto, the joyous quality of a Chopin nocturne, the serene and stately progression of a Bach fugue, the lyrical designs in a Mozart composition for clarinet and strings; the sound of delight in a young boy's voice on catching his first baseball; and, most of all, the expression in the face of someone who loves you—all these are but a small part of a list of wondrous satisfactions that come with the gift of awareness and that nourish even as they heal.

We used to have a lot of fun at the *Saturday Review* in concocting all sorts of absurd notices for the "Personals" section in the back pages of the magazine. The tradition had been started long before my arrival by Christopher Morley and Louis Untermeyer. Readers came to recognize these little spoofs and prodded us to keep them going when we missed an issue or two. Most of the "Personals" notices, of course, were genuine, not a few of them of a lonely-hearts nature. We took pride in the astoundingly large number of families that owed their existence to the postal junctions that came about as a result of the "Personals" columns.

The spoofing notices never intruded into the lonely-hearts category, except for those items that were so outrageous that they could be readily identified. For the most part, the spoofs had to do with computer errors or improbable situations that come up in the course of daily living.

I take special pleasure here in sharing some of the "Personals" that came in the packet from the *Saturday Review* during my hospital stay:

HOW DO YOU KNOW THAT Howard Hughes hasn't remembered you in his will? Our researchers have compiled a master index of 469,000 names listed in the various wills discovered since his death. You may have a bundle waiting for you. SR Box HH.

WE WISH TO APOLOGIZE publicly to the 796 members of the World Stamp Collector's Society who went to Norwalk, Connecticut, instead of Norwalk, California, for our annual convention because of printer's error on invitation. M. G. Stuckey, President, WSCS. SR Box AC.

ARE YOU NO LONGER ABLE TO SPRING out of bed in the morning the way you used to? Our device, attached to your alarm clock, is adapted from an electric cattle prod and can be readily attached to your bedsprings. Quick morning starts guaranteed. Fast-Riser Service. SR Box EC.

NOTICE TO WEARERS OF OUR NEW TOUPEES (Series B-143): Please be sure to cover up in the sun because of chemical factor that causes hair to turn green. PERMATOP Company. SR Box NT.

COMPUTER ERROR has resulted in large supply of electric-powered swivel chairs that make approximately 150 high-speed revolutions per minute automatically as soon as body weight hits the seat. Excellent bargain for people who are nausea-resistant. SR Box SC.

TO THOSE WHO HAVE COME TO THE END OF THEIR ROPES and not yet received our splicing instructions: Please hang on a little longer; we are currently caught up in computer snarl. BINDING TIES, Ltd., WM Box MS.

WONDERING WHAT TO DO if your waterbed freezes this winter? Our trained hamsters (sold in pairs only) skate for hours to the tune of the beautiful "Blue Danube" waltz. Pair sold complete with 33 LP record and 8 extra miniature skates. WM Box D.

NOW SPACE AGE technology makes it possible for you to avoid those frantic runs to the kitchen during the TV commercials. Our combination TV-refrigerator keeps you where the action is. Single unit in gleaming white adds to decor of any living room. WM Box ML.

ARE YOU PREPARED if the President drops in for a surprise visit? Every home should have a record of "Hail to the Chief" at hand to avoid embarrassment. The harmonica arrangement to this inspiring musical tribute is now available. WM Box FW.

LADIES: Tired of lecherous whistles on the street, indiscreet pinches in the subway, ribald remarks from construction workers? My lifelike warts and moles can be instantly attached to nose, chin, and cheeks. Just as easily removed. Write WM, Box RC3.

NOTICE: Due to circumstances beyond our control, the forty-second annual reunion of the Vestal Virgins of America Society will not be held this year. Cynthia P. Cartwright, President.

WAIT! DON'T THROW OUT your warped or rusted railroad tracks, obsolete freight cars, without calling us first. W.M., Maintenance Dept. LIRR.

IF YOU'RE ALLERGIC TO DOGS and still want to protect your hearth and home, we have the perfect solution. Our trained snapping turtles don't shed, bark, or beg to be petted, but will strike fear in the heart of any thug. Our little guardians are trained to lunge for the ankles. Heavy boots also available for members of your household. Write for information. WATCH TURTLES, Inc. WM, Box SK 2.

IMPORTANT NOTICE: If you are one of the hundreds of parachuting enthusiasts who bought our course entitled "Easy Sky Diving in One Fell Swoop," please make the following correction: On page 8, line 7, change "state zip code" to "pull rip cord."

PUBLIC NOTICE: We are withdrawing from circulation the manual *Moth-Training Made Easy*. Reports from readers indicate 15 to 20 percent of trained moths attack their master. No need for alarm; moth bites may be painful but are not poisonous. As soon as our moth experts find out what has gone wrong, we will return your book with further instruction. Moth Specialties Co., New York, N.Y.

DISTRESS SALE: Marginally defective calculators (pocket size). Guaranteed to be 98% accurate. Perfect gift for easygoing students, tax accountants, stockbrokers. WM, Box R.R.

DO YOU REMEMBER the thrill when your first dog "shook paws" with you and your friends? Now you can entertain all your guests simultaneously with one of our trained octopuses. WM Box C.

DO YOU PUT your garbage out at 5:00 A.M. because you are too embarrassed to have your neighbors see how modest your circumstances really are? Now you can hold up your head again. We can supply you with empty imported champagne bottles, beer bottles, caviar containers, pate packages and truffle tins so that those 'garbage snoopers' really have something to talk about. HI-CLASS GARBAGE, INC. WM Box RS.

INSOMNIA A PROBLEM? Fall asleep instantly by listening to our LP recordings of congressional roll-call votes.

Write: Surplus Sound Co., SR/W Box WG.

NOTICE to our Denture Stick-Tite customers: Our scientists have devised a workable release solvent. Consequently, our thanks to wearers who were stuck with old formula. SR/W Box N.S.

WILL PARTY who checked pair of rabbits Grand Central just before Christmas please advise disposition of additional units. Box 114 SR/W.

DON'T BE EMBARRASSED by your parrot's gastronomic imitations. Our researchers have devised a simple food that produces contractions in your parrot's tongue and inhibits rude burplike noises. Burp-Free Parrot Food, Inc. SR Box FB.

Duke University Comprehensive Cancer Center
Durham, North Carolina

[The Comprehensive Cancer Center at Duke University has prepared a list of humorous and other materials available to its patients. This list, so far as I know, is the most far-reaching of its kind—N.C.]

BOOKS

AUTHOR	TITLE
Allen Woody	*Without Feathers*
Baker, Russell	*The Rescue of Miss Yaskell and Other Pipe Dreams*
Bloch, Arthur	*Murphy's Law*
Bloch, Arthur	*Murphy's Law Book Two*
Blount, Roy, Jr.	*Crackers*
Bombeck, Erma	*The Grass Is Always Greener over the Septic Tank*
Boynton, Sandra	*Chocolate: The Consuming Passion*
Breathed, Berke	*Penguin Dreams: And Stranger Things Bloom County*
Buchwald, Art	*You Can Fool All of the People All of the Time*
Buchwald, Art	*I Never Danced at the White House*

Buchwald, Art	*The Bollo Caper*
Burns, George	*Dr. Burns' Prescription for Happiness*
Burns, George	*Dear George*
Camp, Joe	*Oh Heavenly Dog!*
Combs, Ann	*Helter Shelter*
Cuppy, Will	*Decline and Fall of Practically Everybody*
Davis, Jim	*The Fourth Garfield Treasury*
Diller, Phyllis	*The Complete Mother*
Dwyer, Bill	*Dictionary for Yankees*
Ephron, Delia	*How to Eat Like a Child: And Other Lessons in Not Being a Grown-up*
Evans, Greg	*Is it Friday Yet, Luann?*
Fields, W. C.	*I Never Met a Kid I Liked*
Gately, George	*Heathcliff Smooth Sailing*
Greenburg, Dan	*How to Make Yourself Miserable*
Grizzard, Lewis	*Don't Sit Under the Grits Tree with Anyone Else but Me*
Grizzard, Lewis	*Shoot Low, Boys—They're Ridin' Shetland Ponies*
Hewlett, John	*The Blarney Stone*
Horn, Maurice	*Comics of the American West*
Keillor, Garrison	*Happy to Be Here*
Kerr, Jean	*Please Don't Eat the Daisies*
Larson, Gary	*The Far Side Gallery Two*
MacNelly, Jeff	*The Greatest Shoe on Earth*
Millar, Jeff, and Hinds, Bill	*Tank McNamara*
Ohman, Jack	*Drawing Conclusions: A Collection of Political Cartoons*
Pizzuto, John	*The Great Wall Street Joke Book*
Powell, Dwane	*The Reagan Chronicles*
Schulman, Max	*Rally Round the Flag, Boys!*
Smith, Wes	*Welcome to the Real World*
Viorst, Judith	*It's Hard to Be Hip Over Thirty and Other Tragedies of Married Life*
Wilde, Larry	*The Official Executive's Joke Book*
Wilder, Roy, Jr.	*You All Spoken Here*
Winters, Jonathan	*Mouse Breath Conformity and Other Social Ills*

AUDIOCASSETTES

PERFORMER	TITLE
Anonymous	*Bloopers*
Clower, Jerry	*The Ambassador of Goodwill*
Clower, Jerry	*Live from the Stage of the Grand Ole Opry*
Clower, Jerry	*Top Gum*
Clower, Jerry	*The One and Only*
Clower, Jerry	*Runaway Truck*
Clower, Jerry	*Live in Picayune*
Cosby, Bill	*The Best of Bill Cosby*
Cosby, Bill	*Is a Very Funny Fellow, Right!*
Cosby, Bill	*Wonderfulness*
Cosby, Bill	*Inside the Mind Of*
Cosby, Bill	*200 M.P.H.*
Dangerfield, Rodney	*I Don't Get No Respect*
Fields, W. C.	*The Best of W. C. Fields*
Gardner, Gerald	*All the President's Wits*
Marx, Groucho	*The Works*
Keillor, Garrison	*News from Lake Wobegon—Fall*
Keillor, Garrison	*News from Lake Wobegon—Spring*
Keillor, Garrison	*News from Lake Wobegon—Summer*
Keillor, Garrison	*News from Lake Wobegon—Winter*
Nash, Ogden	*Ogden Nash Reads*
Rivers, Joan	*What Becomes a Semi-legend Most?*
Stevens, Ray	*Crackin' Up*
Stevens, Ray	*Greatest Hits*
Stevens, Ray	*Surely you Joust*
Stevens, Ray	*He Thinks He's Ray Stevens*

VIDEOCASSETTES

ADVENTURE

Cool Hand Luke
"Crocodile" Dundee
High Road to China
Jake Speed
Jeremiah Johnson
Jewel of the Nile
Patton
Raiders of the Lost Ark

Raise the Titanic!
Rocky III
Romancing the Stone
Silverado
Superman
Top Gun

CLASSICS

The Bridge on the River Kwai
Casablanca
From Here to Eternity
On Golden Pond
To Kill a Mockingbird

HUMOR/COMEDY

Airplane!
All of Me
Back to the Future
Blazing Saddles
The Films of Laurel and Hardy
Making Mr. Right
The Making of the Stooges
Privates on Parade
The Return of the Pink Panther
Silverado
Some Like It Hot
Volunteers

MUSICALS

The Jolson Story
Jolson Sings Again
The Sound of Music
42nd Street
White Nights
Rick Springfield

SCIENCE FICTION

Star Wars
Star Trek IV: The Voyage Home

WESTERNS

The Alamo
Shane
True Grit
Hang 'em High

DOCUMENTARIES

National Geographic—Iceland

MISCELLANEOUS

Scenes at Duke

11

MEDICAL MISTAKES

Medical science has made enormous progress. Even so, it is constantly coping with error—not only of theory but of practice. But mistakes have a way of concealing themselves or, indeed, of becoming institutionalized. A classic example is afforded by the treatment of George Washington in his final illness.

On the morning of December 12, 1799, Washington mounted his horse and rode out to his farm. In his pocket was a plan for the coming year about developing the land, rotating the crops, and taking an inventory of the livestock.

The weather turned cold and wet. By early afternoon, sleet alternated with heavy snow. The general returned home at about 3:00 P.M., his bare head frosted with ice. By evening, chills and sniffles set in. Early the next morning, he told Martha Washington he had a sore throat but thought it of no consequence and went about his business. Despite a developing hoarseness, he went outside late in the afternoon to mark trees for cutting. When he came back to the house his cold had worsened. During the night, his breathing became labored; sleeping was fretful. At about 3:00 A.M. he complained to his wife about the fact of deepening illness. At daybreak, Mrs. Washington sent for Dr. Craik. Even before the doctor arrived, the general ordered one of his employees to begin the standard treatment of the time—drawing off blood.

General Washington was a good friend of Dr. Benjamin Rush, one of the most eminent physicians of his time and a strong advocate of bleeding to counteract illness. Washington was familiar with the procedure and supervised his own bloodletting, saying that the initial incision was not large enough. As the bloodletting progressed, Martha expressed concern at the amount of blood being drawn. The general reassured her and instructed the servant to continue with the procedure.

Martha directed the preparation of a concoction designed to soothe the throat—vinegar, molasses, and butter—but the general's throat was so sore by this time that he was unable to get anything down.

A strip of warm flannel was wrapped around the general's neck and his feet were bathed in warm water—all to no avail. The delay in Dr. Craik's arrival prompted Mrs. Washington to send for Dr. Brown. However, Dr. Craik arrived before Dr. Brown and promptly ordered continuation of the bloodletting. He then asked that the general be given a strong gargle for his throat. In addition, a steam inhalant with vinegar was prepared. The general tried to gargle but almost suffocated.

Dr. Brown not having arrived, Dr. Craik felt in need of expert assistance. A messenger was dispatched for Dr. Dick, who lived nearby. Both Dr. Brown and Dr. Dick arrived at about 3:00 P.M. The three physicians had a consultation and it was agreed that the general was rapidly weakening and that further bleeding was urgently necessary.

At 4:30 P.M. Mrs. Washington was summoned to the bedside and the general requested that she go to his desk and find his two wills. She did so. The general identified the valid will and ordered the other to be burned.

Then he said he knew he was dying. He summoned his secretary and instructed him in the settlement of all accounts. He also requested that his letters and military papers be put in final order. He asked his secretary if he could think of anything that was left undone and was assured that everything had been attended to.

Meanwhile, the bleeding was continued. The general whispered that the end was near.

"I die hard," he told Dr. Craik, "but I am not afraid to go. I believed from my first attack that I would not survive it. My breath cannot last long."

A short while later, he told the doctor: "I thank you for your attentions; but I pray you to take no more trouble about me. Let me go quietly."

Mrs. Washington noticed that the general was very still.

"Is he gone?" she asked.

No one said anything. The secretary slowly raised his hand in assent. Death came Saturday evening, December 14, 1799.

Today's physicians, reviewing the treatment of General Washington, would be appalled by the bleeding, which was strenuously continued even as the illness worsened, and which undoubtedly hastened his death. There is no record of the precise amount of Washington's blood loss, but the indications were that at least four bleedings were taken, which could have totaled two quarts. Even for a healthy person, such a loss would not be without risk.

Bleeding is only one of the many treatments applied over the centuries that are recognized today as harmful or even barbaric. Patients have been subjected to all sorts of improbable and foul-tasting prescriptions, all the way from animal urine and excrement to snake oil and shavings of goat's horns. As late as the beginning of the twentieth century, bleeding, forced retching, and purging were still being used by many doctors.

The ancient Egyptians, according to Herodotus, made a standard practice of forced total bowel evacuation for three consecutive days each month. Other cultures have been similarly preoccupied with the condition of the human colon. Dr. Edward C. Lambert, to whom I am indebted for many of the facts presented in this chapter, wrote a valuable book titled *Modern Medical Mistakes*, which details this obsession with the digestive tract. He cites Molière's *Imaginary Invalid*, in which a doctor's bill included an item for "a small injection preparatory, insinuation, and emollient, to lubricate, loosen, and stimulate the gentleman's bowels . . . to flush, irrigate, and thoroughly clean out the gentleman's lower intestine." He also quotes Jonathan

Swift's description of Gulliver's conversation with the Lilliputians about English habits: "They take in . . . a medicine equally annoying and disgustful to the bowels, which, relaxing the belly, drives down all before it; and this they call a purge."

The purge continued to dominate treatment through the early years of the twentieth century. Lambert tells us that Elie Metchnikoff, one of the great names in Russian medical science, believed that the large intestine was an accident of nature that, like the appendix, should be surgically removed. A character in Aldous Huxley's *Eyeless in Gaza* asks: "How can you expect to think in anything but a negative way when you've got chronic intestinal poisoning?"

In the early years of the twentieth century, "colon laundries" sprang up on opposite sides of the Atlantic. The entire world seemed bound together by a preoccupation with constipation, real or imagined.

Little by little, however, the anatomical focus of attention shifted. Now it was the appendix that was identified as the source of all evil and had to come out. Then the thymus gland became an object for radiation, especially for infants. This craze, fortunately for the immune system, quickly ran its course, followed by the ubiquitous tonsils, a blazing symbol of the rapidly exploding new consciousness about disease germs. Identified as a gathering place for tiny enemies of humankind, the tonsils were a prime candidate for excisions. Anyone who got as far as puberty and who still had tonsils was regarded as a victim of parental neglect.

Up and up the scalpel, scissors, or extractor moved. Now it was the teeth that were seen as baleful agents, havens for insidious infections that could lead to all sorts of illnesses, arthritis high among them. These pervasive threats could be conveniently combated by extracting all the teeth. Many thousands of people in their twenties or thirties had to spend the rest of their lives coping with manufactured dentures.

Gradually, however, a collective wisdom began to emerge and it was recognized that the basic design of the human body was not all that faulty or capricious; that not everything for which a specific purpose could not be found had to come out or be radiated; that most people could coexist quite satisfactorily with their own organs, including the appendix, the thymus gland, and the tonsils. Infected teeth could be treated and saved. Bacteria lost their terror as ultimate culprits to be relentlessly hunted down and exterminated. It was even discovered that most of these bacteria played a useful role in the total economy of the human organism. People gained a new respect for what Dr. Walter Cannon called the "wisdom of the body."

Lest it be thought, however, that the human species has ascended to a plateau of perfection today, totally free of error in the treatment of human disease, it may be useful to scrutinize evidence of recent fallibility.

Some things stand out immediately. Until very recently, people with heart attacks were kept in bed for a month or more following a myocardial infarction. Strictures against exercise were so severe that, during the first few days, patients were protected against even the minimal exertion of brushing their teeth.

Today's cardiologists recognize that such treatment is not only undesirable but harmful. They know that even a damaged heart has a capacity for exercise that needs to be carefully expanded. Patients are encouraged to get out of bed only a few days after a myocardial infarction. They are also encouraged to follow a life-style emphasizing reduction of emotional stress, improved nutrition, and graduated exercise. Few things are more important for heart-attack survivors than to avoid becoming psychological cripples, with panic as a constant companion.

Ulcers are another illness for which treatment has been drastically changed just in the past few years. For countless decades, it was considered necessary for ulcer sufferers to stay on a bland diet—puréed foods, milk products, mashed potatoes, etc. Today it is recognized that high-fiber foods represent an effective antiulcer diet.

Bypass heart surgery is one of the prime surgical advances of the twentieth century and may represent the difference between life and death in some cases. Also, the improvement in the quality of life for many bypass patients must be recognized. However, many thousands of such complex operations have been performed that were not necessary, according to a study carried out by the National Heart, Lung, and Blood Institute. During the 1980s, an average of 230,000 bypass operations were performed annually. The health status of 780 heart disease patients over a period of five years (one group received bypass operations; the other medical therapy) was carefully followed. The study showed that the five-year survival and heart attack rates were the same for those who received medical therapy and those having bypass surgery. A similar European study confirmed the finding that both medical and surgical treatment resulted in virtually identical heart attack rates after five years. Since bypass surgery hospital costs run up to $30,000—in addition to surgeon's fees of $5,000 or more—the economic aspects and consequences cannot be overlooked.

There is a tendency to put the entire blame on cardiac surgeons for the thousands of unnecessary bypass operations performed each year. What is generally not recognized is that many hospitals put pressure on surgeons to get patients into the operating rooms. In many cases, the hospitals are responding to pressures from third-party payers—government and private insurance agencies. Hospitals have to cope with "DRGs" (designated related groupings). What this means is that the insurers have drawn up their own scale for reimbursement to hospitals. A designated disease carries with it a specific number of allowable hospital days, regardless of individual variations. These regulations have been modified several times in response to substantial abuses.

One of the ironic aspects of the third-party payer system is that, in some places, it has actually resulted in incentives for performing bypass surgery. In California, for example, a hospital has to perform a minimum number of bypass operations in order to qualify for state payments. Hospitals have been penalized, in effect, for not coming up to their quotas. Little wonder, as stated above, that the study conducted by the National Heart, Lung, and Blood Institute revealed that at least 25 percent of all bypass operations were unjustified.

Some surgical procedures have been found to be faulty. In the 1950s, a new form of treatment for broken bones came into favor. A Soviet surgeon reported on his success in treating fractures by gluing the bone together. He used a resin with strong adhesive qualities. This technique was refined in Australia by an orthopedic surgeon who bonded fractures with ethoxyline resins, which apparently had even greater adhesive power than the resin used by the Soviet orthopedist. What made the Australian report seem so attractive was that it was successfully used on several patients whose fractures had failed to yield to conventional splints.

The next entry came from Philadelphia, where an orthopedist reported having successfully used a plastic called "Ostamer" for connecting broken bones. After only three days, the patient was walking without a cane or crutches.

Unfortunately, most of the "successful" results did not stand the test of time. Long-term failures began to be reported in the medical press. Finally, it was discovered that many of the bonding chemicals did not dissolve and that they carried all the liabilities of a foreign body. Errors both in concept and application became increasingly evident and the technique was abandoned.

In 1976 a congressional subcommittee on Oversight and Investigations reported the results of a study of surgery lacking sufficient justification. Among their conclusions: American doctors performed 2.4 million unnecessary operations in 1974, which resulted in 11,900 deaths and a price tag of $4 billion.

High on the list of surgical excesses are hysterectomies. The National Institute of Health Statistics reported that of the 750,000 hysterectomies performed annually, 22 percent are unjustified and, moreover, only 10 percent of all the hysterectomy operations were uncontestably warranted.

A joint study by the departments of Medicine and Public Health at UCLA and the Rand Corporation in Santa Monica, California, evaluated the appropriateness of the carotid endarterectomy—a procedure commonly used in the prevention of stroke, yet controversial because of its questionable efficacy and high complication rate—of which 107,000 were performed in 1985. The researchers undertook a rigorous evaluation of the records of 1,302 elderly carotid endarterectomy patients in 1981. Utilizing the expertise of a panel of nationally distinguished experts, the investigators concluded that only one-third of the surgeries done were warranted, one-third were highly inappropriate, and one-third were done with equivocal indications.

The same investigators also judged that out of 386 coronary artery bypass cases they evaluated between 1979 and 1982, 30 percent of the surgeries were performed for questionable reasons and 14 percent were clearly inappropriate. They also classified as inappropriate 17 percent of 1,677 coronary angiography and 17 percent of 1,585 upper gastrointestinal tract endoscopy examinations conducted in 1981.

Yet another demonstration of unjustified surgery in some cases is that of cardiac pacemaker implantations. One in 500 Americans has received a permanent cardiac pacemaker at an average cost of $12,000 per implant. An estimated 120,000 implantations were performed yearly in the 1980s. The U.S. Senate Special Subcommittee on Aging found that up to half of Medicare-covered costs for pacemaker implants were unwarranted because the surgeries carried questionable medical justification. Dr. Allan M. Greenspan and his colleagues at the Albert Einstein Medical Center in Philadelphia scrutinized the records of 382 elderly cardiac pacemaker recipients. The researchers found 20 percent of the implantations to have been completely unjustified and done based on a misdiagnosis of information, and found 36 percent of the procedures to have been done with incomplete diagnostic evaluations or inadequate documentation.

Cosmetic surgery has come into increasing favor in recent years. Face-lifts account for most such procedures but, increasingly, surgery is now being used as a shortcut to weight reduction, breast enlargement or reduction, hair replacement, etc. All surgery, however, involves risks—infection and nerve damage among them. The anesthesia that accompanies surgery can represent even greater risk since it can have profound effects on heart function.

One of the patients I was asked to see was an artist in her mid-forties. She had anguished over her near-obesity. The various diets she had tried were unsatisfactory for one reason or another. Then she read about a quick new way of reducing oversized waistlines. This new surgical procedure, lipectomy, would simply suck out abdominal fat until a desired reduction was achieved.

In this particular case, however, the woman emerged from the surgery with arm paralysis. The surgeons said that the limitation of movement would gradually pass. After a month, however, the paralysis was unchanged. She slid into a deep depression, incurring yet other health risks. I discussed the matter with the surgeon, who was confident that mobility would gradually return and he felt that her own confidence was essential not just to recovery but to her quality of life and to her psychological state. He called attention to the fact that the artist had signed an informed-consent paper in which the various hazards of the surgery were identified. In her eagerness to have a quick fix to her surplus poundage, the woman tended to regard the document as a routine legal measure. My main concern was connected to her need to avert panic and depression, which could actually retard the process of recovery. I urged her to accept psychotherapy, especially since she felt she was being punished for her vanity. Meanwhile, she was not barred from pursuing her interest in art. She had ideas that might be put into writing. Activity, no less than confidence, is deeply connected to the environment of restoration.

In reflecting on this case, I realized again how precarious is the quick fix. We tend to turn to pills for instant correction of almost any conceivable problem—all the way from headaches to sleeplessness. Painkilling pills have become a substitute for sensible attention to the causes of pain. We expect the surgeon's knife or the prescription pad to replace the personal discipline required to maintain good health. We even learn speed-reading, trying to negotiate a page in reduced time but also depriving ourselves of the richness of imagery and thought that good writing offers those who know how to hover over a carefully crafted sentence. In the act of getting from here to there in half the time we reduce our perspective and lose our options.

Good health is a serious business. Like life itself, it has to be worked at and it takes on added meaning with effort.

One of the most conspicuous examples of modern medical mistakes, of course, has been the widespread use of X-ray equipment. Until comparatively recently, medical examinations would include routine use of X-ray procedures and, to a lesser extent, fluoroscopy. Fluoroscopy is to the X ray what the motion picture is to the still photograph. Fluoroscopy has the advantage of enabling the examiner to have a continuing internal view of body processes, but it delivers many times the radiation of a single X-ray plate.

X rays, whether by "stills" or continuous viewing, are not innocuous. They are not merely a strong light that enables the viewer to see through to the insides of things. They are active and specific impulses that penetrate human tissue, registering on a plate on the opposite side. Just in the act of piercing tissue they can create change or do damage. The amount of radiation in a single exposure is very low and the amount of damage is so small as to be minimal. But the effects of radiation are cumulative and the body keeps track of the total score. The ability of human beings to absorb radiation is not infinite. Limits tend to vary with individuals. Limits also vary with other factors, such as the altitude of one's home or workplace, exposure to solar radiation, or industrial radiation, etc. Today's physicians are not unaware of these hazards and tend to be more conservative than their immediate predecessors, but it is a mistake to assume that excessive or unnecessary use of X-ray equipment has been totally eliminated.

A procedure that was widely believed to be efficacious was the use of insulin shock in the treatment of schizophrenia. In 1939, when I was a young medical reporter, I attended a consulting session at the Hastings Hillside Hospital in Westchester County, New York. About a dozen psychiatrists would listen to presentations by the hospital physicians of individual cases. After discussion of the presentation, the individual patients would be brought into the room for questioning by the consulting psychiatrists. There seemed to be an effort by the group to arrive at a consensus as to whether the proper diagnosis was either schizophrenia or manic-depressive disease. If the former, insulin shock was indicated; if the latter, electroconvulsive (electric shock) therapy. Several years later, the procedures were switched in some places: electric shock for schizophrenia and insulin shock for manic-depressive disease. Later, it was reversed again.

What puzzled me deeply at the time, and still does, was the insistence that mental disease had to be either schizophrenia or manic depression. It seemed logical to suppose that some patients could be suffering from a combination of both illnesses in varying degrees, or belong to an entirely different category of mental illness, as yet not classified. Wasn't it possible that the individual placed his or her own stamp on the illness, representing a manifestation of biological deficiencies or emotional perturbations or life-style abuses? Dr. Jolly West, with whom I have had long discussions, has emphasized the rapidly evolving nature of psychiatric theory and practice. He regards the Hastings Hillside Hospital consultations as examples of early approaches long since discarded. Even so, it may be useful to consider the history of insulin shock.

In the early 1930s, Dr. Manfred Sakel, a German neuropsychiatrist, reported his experiences in putting schizophrenia patients into forced coma through injection of insulin. He arrived at this theory through accident. A diabetic morphine addict was given an accidental overdose of insulin, following which the patient's confused mental state appeared to clear up. It was theorized that the insulin, by forcing sugar out of the blood and into the cells, brought about improved metabolic activity.

Insulin shock came into general use as the treatment of choice in schizophrenia despite the evidence that the therapy was accompanied by a wide range of adverse effects, including irregular heartbeat, sweating, severe digestive problems, nausea, stupor, and, finally, coma and sometimes convulsions—all of which are to be expected with a sudden and severe fall in the level of blood sugar.

How, then, to explain the fact that many patients treated by insulin experienced improvement? Psychiatrists who have studied the question tend to agree that some of the improvement was probably attributable to the extra attention given the patients because of the severity of the procedure. The placebo effect may play its part, even in patients with severe mental illness. It may also be that the shock mobilized certain latent psychobiological coping mechanisms still unknown.

Electroconvulsive therapy (ECT) succeeded insulin shock as a treatment for manic-depressive and schizophrenic patients during the 1940s and 1950s. It was widely used even though the reasons for its benefits were not understood. Perhaps the same reasoning may apply here as has been used in relation to insulin shock. With the development of numerous effective antipsychotic medications since the 1960s, electric shock has fallen into increasing disfavor. Nevertheless, it is still used in certain cases when medication fails, or is contraindicated, and nobody knows why it sometimes works. One thing, however, is certain. The original rationale for convulsive therapy—that epileptics are immune from schizophrenia—was a mistake.

•

Mistakes, misconceptions, or failures are not confined to types of treatment or procedures. Medical products or preparations cannot be excluded from the negative inventory. Not long ago, for example, babies in hospitals would be routinely washed with soap containing hexachlorophene or would be powdered by talcum containing the chemical. Then it was discovered that the talcum or soap, when discontinued, opened the gates to staphylococcal infection. It was also discovered that frequent use of the preparation could damage the central nervous system. The Food and Drug Administration and the Committee on Fetus and Newborn of the American Academy of Pediatricians issued a statement calling attention to the dangers of the preparation.

Human skin is not impervious to powerful chemicals. Cardiac patients are able to absorb appropriate amounts of nitroglycerin in paste form when attached in a patch to the inner arm. Other medications, absorbed through the skin, can create an immediate taste sensation. Such being the case, it is not surprising that a powerful chemical such as hexachlorophene, when used frequently, could be harmful. If used at all these days, the preparation is carefully supervised and is rinsed off thoroughly with water.

Medications involving the use of mercury have also come under careful scrutiny. In Great Britain, young children whose teeth had been cleaned by mercurous chloridex developed severe pains in the extremities. In the United States, physicians called attention to the dangers of the preparation but it was thirty years before the FDA was able to compel drug manufacturers to eliminate mercury from tooth powders.

Around the turn of the century, a drug called amidopyrine came into use both in Europe and the United States. It was hailed as a super pain reliever. However, people began to be afflicted by inflammation of mouth and throat, fever, and all sorts of infections. One of the characteristics of the illness was loss of circulating white blood cells. An increasing number of deaths was traced to the drug. Autopsies in these cases produced stark evidence of toxicity. Some medical investigators theorized that a benzene ring in the drug was the cause of the illness.

Weight-reduction pills have been implicated as causes of serious illness. In 1931, the drug dinitrophenol, also used as an ingredient in explosives, was used for combating overweight. The result in some cases was not just loss of weight but loss of life. Though the drug was widely advertised as being completely safe, an increasing number of young women began to develop bone-marrow deficiency. Then came reports of cataracts as a result of use of the pills. Finally, in 1935, German health officials issued warnings against the drug. Canada banned the drug altogether. A year later Great Britain removed the drug from the approved list of nonprescription medications. The United States didn't take the drug off the market until 1938.

After World War II, one of the drugs prescribed by some physicians for weight reduction was amphetamines, which had demonstrated value for certain clearly defined illnesses such as sleeping sickness. As a weight reducer, however, it was not without hazards. The drug did succeed in shrinking the appetite. It did other things. It led in some cases to anorexia, especially in teenagers and young women. It did damage to the nervous system. It caused aberrations and mental symptoms that mimicked schizophrenia. Any number of young women were misdiagnosed and sent off to mental institutions, leading in some cases to feelings of severe abandonment and serious nervous disorders that took years to correct. Amphetamines, like many other drugs, can have profound value when appropriately used.

Hormone drugs are not without risk. Dr. Lambert reported in his book the fact of malformations in children following the use of male hormones by mothers during pregnancies, in an effort to reduce the risk of miscarriages. The procedure had been thought to be completely safe, then investigators at Johns Hopkins University Hospital made correlations between the use of the hormones during pregnancy and the birth of female babies with external genitals and other male features. Further investigation identified some 600 female babies who had masculinized genitalia. Despite this evidence, the manufacturer continued for some time to advertise the product as being safe during pregnancy.

The most publicized instance of harm caused by the use of drugs during pregnancy, of course, concerns thalidomide. It came into wide favor abroad during the mid-1950s as a combined pain reliever, tension reducer, and sleep inducer. It was advertised as being completely nontoxic. After five or six years of widespread use, cases began to be reported of devastating effects, some of which manifested themselves long after the drug was discontinued. Children of mothers who took thalidomide during pregnancy had deformations or other abnormalities. Some babies were born with flipperlike arms and legs. Cardiac abnormalities began to turn up. Other health problems did not reveal themselves until the children were fully grown. In West Germany alone more than 6,000 children were crippled or deformed. In the United States, use of the drug was comparatively small. Thanks to the knowledgeable and vigilant action of Dr. Francis Kelsey of the FDA, the drug was not approved for sale in the U.S. Thousands of children of American mothers who had lived in West Germany at the time, or who had acquired the drug in one way or another, have been affected and have joined in lawsuits against the manufacturers.

A current example of mismedication concerns the use of quinines for all sorts of fevers other than those caused by malaria, in which they have a demonstrated efficacy. "Bromo-quinine" tablets are widely used to relieve colds but no scientific evidence exists to justify such usage. Most colds run their course after a few days—a fact that undoubtedly works to the credit of the pills. Bromides are not harmless but are widely dispensed in such over-the-counter products as Bromo Seltzer. Similarly, lithium salts are sometimes used as substitutes for regular salt but can have deleterious effects on cardiac function.

The sheer repetition of these and other horrors might make it appear that *all* medications ought to be avoided under *all* circumstances. But certain medications can be lifesaving. The wise use of antibiotics has brought otherwise lethal diseases under control—diphtheria, tuberculosis, syphilis, gonorrhea, bacterial endocarditis among them. Problems arise largely when medications are overprescribed or when they are not properly monitored. Unnecessarily prolonged use, disregard for their effect in combination with other medications, failure to test the individual patient for possible inability to tolerate a certain drug or for drugs in general—all these hazards call for careful and consistent monitoring. The wise physician is always aware of contraindications as reported in the *Physicians' Desk Reference* or other volumes such as *Hazards of Medication.* The wise physician, too, checks with patients constantly for evidence of side effects. Problems of addiction with certain drugs—tranquilizers and painkillers especially—are marked out for special surveillance by the wise physician.

Even as the wise physician takes full advantage of the armamentarium available to him, he never misses the opportunity to educate the patient to the truth that drugs aren't always necessary and that the human body is its own best drugstore for most symptoms.

Hypnotism may be indicated in combination with drugs—or where certain drugs have proved to be ineffective. Here we have working proof that the power of suggestion, which is a mental process, can have biological effects. In a sense, hypnotism offered valuable clues to the questions behind the obsession that brought me to UCLA.

12

MESMER, HYPNOTISM, AND THE POWERS OF MIND

Knowing of my interest in the way the mind can create biological change, Jolly West invited me to his lecture on Franz Anton Mesmer, who was able to bring about some remarkable reversals of illness through a technique now called hypnotism. That lecture started me off on an exploration into the life and times of the man whose name was to pass into the language as a noun and a verb (*mesmerism, mesmerize*).

On March 12, 1784, a Royal Commission of Inquiry was appointed by King Louis XVI of France to investigate the theories and practices of Dr. Mesmer, whose unorthodox methods of treating patients had turned Paris upside down with excitement and, inevitably, had outraged the medical profession. To fill the post of chairman of the Royal Commission, the king appointed Benjamin Franklin, American minister to France, renowned and respected throughout the civilized world.

The announcement of the commission hit Paris like a bombshell and produced a great commotion. Mesmer had even then become a legendary figure because of his amazing "cures." He was the darling of high Parisian society. Marie Antoinette, the Marquis de Lafayette, and Madame du Barry were among his most enthusiastic patients. Mesmer's "cures" through his system of "animal magnetism" were a favorite conversation piece at fashionable dinner tables.

What gave high drama to the investigation was the prospect of an epic confrontation between its central figures. Mesmer had captured the imagination, Franklin the admiration, of large numbers of Frenchmen. Mesmer's name may have been on their lips, but Franklin's statuette was on their mantelpieces. Mesmer was mystical, melodramatic, flamboyant; Franklin was open, disarming, wise. To many in the science academy, Mesmer was a throwback to earlier times of black magic and pervasive superstitions when religion and medicine frequently came off the same arcane spool.

By contrast, Franklin was recognized as a symbol of the new enlightenment, a period in which people could throw off old answers and ask questions about the nature of life and human rights. A person's religious beliefs were no longer to be mandated by the state but were to be regarded as supremely personal matters. It was felt that what was most important about the search for Truth was not just the nature of truth but the process of reasoning that might lead to it. And what was most significant in science was the scientific method—a way of examining facts and conducting experiments that could be replicated by others. Franklin enjoyed a place of honor in the contemporary rationalist pantheon, along with Voltaire, Diderot, Lavoisier, Rousseau, Condorcet, Priestley, Jefferson, Paine, and John Adams.

Louis XVI had other reasons for appointing Franklin as chairman. The king was caught between competing pressures—on one side, the royal scientific academies, which regarded Mesmer as a charlatan and swindler; on the other, the general public, which saw Mesmer as a master healer and a challenge to authority. Franklin was probably the only person whose own popularity was strong enough to stand against Mesmer and relieve the pressures against the king. As a scientist himself, Franklin was readily accepted by the academicians. As a human being, no one was more loved by the populace.

What lent additional spice in the public mind to the confrontation was the fact that Mesmer had been accused of operating a sex parlor under the guise of a medical clinic. Most of his patients were women. It was said that he would deliberately arouse his female patients to a point where they would gasp rhythmically and reach culminations that had nothing to do with medical treatment. Mesmer's "touching" techniques were denounced as being little more than intimate provocations. He would sit directly opposite the woman, thighs pressing against each other, and, not infrequently, his hands would make "passes" or manipulate the regions of the body nearest the affected organs. Mesmer was said to believe that these organs called for stimulation, a procedure that generally went unopposed by many of his female patients. The writhings and moanings and wild gestures in the final stages of Mesmer's group treatments were interpreted by other physicians as being closer to sexual climaxes than emotional peaks clearly connected to scientific therapy.

Mesmer would appear at these sessions dressed in a long lavender-colored silk gown, waving a magnetic wand over the patients, who were not always fully dressed and who were connected to each other with a rope. These facts were not overlooked by those French physicians who objected to the activities in Mesmer's lavishly decorated treatment center.

Another serious charge lodged against Mesmer by the medical establishment was that he deliberately created an air of mystery about his system in order to create a market for selling his "secrets" for profit. Mesmer and his associate, Dr. Charles D'Eslon, organized a "Society of Harmony." For very substantial fees, people could join the society and be the beneficiaries of Mesmer's secrets. Mesmer claimed that he could magnetize anything—people, dogs, trees, trays, pebbles, wool, glass, paper, water. He said he could also, under certain circumstances, transfer this power to others. The notion that hundreds, perhaps thousands, of wand-waving laymen might run around France treating people for their illnesses with "animal" magnetism was hardly soothing to fully accredited physicians. In toto, it was asserted that the Society of Harmony was a blatant swindle, out of which Mesmer and D'Eslon were making a fortune.

The king held off action on Mesmer as long as he could, fortified by the pro-Mesmer urgings of Marie Antoinette. But counterpressure steadily mounted from the scientific academies that claimed that Mesmer's basic ideas about animal magnetism were a plagiarized concoction. He was accused of having stolen the essentials of his theory from various other theorizers or practitioners, including the medieval physician Augustus Philippus Aureolus Theophrastus Bombastus von Hohenheim, better known historically as Paracelsus.

Mesmer used the term *animal magnetism* to describe what he claimed was a universal process and basic life force. The main feature of this force, Mesmer contended, was a highly rarefied fluid that was present in the universe, keeping the planets and other heavenly bodies in their places or orbits. In human beings, it provided fundamental energy and vital balance. When the fluid dropped below a certain level the body became susceptible to all sorts of disorders or derangements. He asserted his magnetic techniques had the effect of raising the vital fluid to its proper level, thus restoring an inner "harmony" or health.

When the investigating commission held its first meeting at Franklin's home in Passy, then a fashionable suburb of Paris, the seventy-eight-year-old American world citizen could look around at an eminent group of scientists. Among its members were Dr. Antoine-Laurent Lavoisier, who was to win a reputation as the father of modern chemistry; Dr. Antoine-Laurent de Jussieu, the botanist; Dr. Joseph-Ignace Guillotin, the prominent physician whose name was later to be associated with the automatic decapitation device that replaced what he termed was the "barbarous" hand-held ax; and Jean-Sylvain Bailly, astronomer and statesman, who was to become mayor of Paris and who was to lose his head on the guillotine.

The commission began its work in late March and continued its sessions at irregular intervals until early August. Most of the meetings were held in Passy, sparing the gout-plagued Franklin from the need to travel elsewhere. Members of the commission visited Mesmer's salon without Franklin and observed animal magnetism in practice. Some of Mesmer's patients were brought out to Passy, being "treated" in Franklin's library or garden. On one occasion, a tree in the garden was "magnetized" and a child was "cured" of his affliction as he approached the tree.

Franz (in some places listed as Friedrich) Anton Mesmer was born on May 23, 1734, in Iznang, a small German village near Lake Constance. His father was a gamekeeper, his mother the daughter of a locksmith. He was one of nine children and was brought up in modest circumstances. His education had a strong religious orientation. At the age of fifteen he enrolled in a Jesuit college in Bavaria, transferring after three years to the University of Ingolstadt with the intention of becoming a priest. In his sixth year at Ingolstadt, he decided to leave theology for law and enrolled at the University of Vienna. This particular decision lasted little more than a year when he changed from law to medicine at the same university, receiving his doctorate in that discipline at the age of thirty-one.

Mesmer used the term *animal gravity* in his paper to describe the effects of the universal environment on human beings. In this way, he foreshadowed his use of the term *animal magnetism* that was to be identified so closely with his career.

Mesmer's adjective, *animal*, applied primarily to human beings, though he did not exclude other living creatures. His purpose was to make a distinction between mineral forces, of which the magnet is the prime example, and the magnetic forces that existed inside living beings. As his ideas in this direction evolved, he put his emphasis on the magnetic fluid—so "subtle" as to be almost vaporous.

Not long after receiving his medical degree, Mesmer experienced a dramatic change in his personal fortunes. He cultivated the acquaintance of leading composers and musicians, Gluck and Haydn among them. He also became a close friend of Leopold Mozart, whose young son, Wolfgang Amadeus, was electrifying concertgoers throughout Europe with his compositions and operas. Mesmer, with his new wealth, became a patron of the young Mozart. Wolfgang Amadeus gave the premier performance of his opera *Bastien und Bastienne*, which Mesmer had commissioned, in Mesmer's garden at 261 Landstrasse. Mozart also referred by name to Mesmer as the magnetizer in *Così fan tutte*.

Despite slowly mounting criticism from other physicians, Mesmer's Viennese practice grew and indeed flourished, his patients including prominent members of the nobility. What finally defeated Mesmer in Vienna was not so much the opposition from within his profession as it was the public charges that he was using the privilege of his practice to compromise young ladies. In particular, one of his patients was a beautiful and talented eighteen-year-old blind pianist, Maria Theresa von Paradis, a protégée of the empress Maria Theresa. Young Mozart, deeply moved by the genius and the courage of the young lady, had composed a concerto in her honor.

Appalled by the failure of conventional medicine to restore the girl's sight, Mesmer persuaded her parents to entrust her to his care. Using animal magnetism, augmented by specially designed mirrors, and drawing upon his "inner magnetism" for raising Maria's level of the vital life fluid, he claimed to succeed in partially restoring her sight. Because of the frequency of the treatments he said were necessary, the parents acquiesced to his request that Maria be permitted to live in the Mesmer household. He had diagnosed her blindness as amaurosis, a functional rather than organic impairment of the optic nerves, and therefore susceptible to his treatment.

The result not just of the Paradis case but of other complaints was that the Viennese medical faculty had all the ammunition it needed against Mesmer. A commission was appointed to look into every aspect of his practice. The finding was unambiguous. His cures, such as they were, were attributed to the effects of the patients' own imaginations. The comment on the Paradis case was little short of devastating. Mesmer's system was labeled fraudulent. He was ousted from the medical fraternity and barred from further practice.

Mesmer left Vienna. He also left his wife, a circumstance that did not surprise friends to whom he had confided that he thought her stupid and a bore.

His next dramatic entrance on the medical stage was in Paris, where he was able to surpass even the considerable early success he had enjoyed in Vienna. His ebullience, charm, self-confidence, and showmanship gave no hint of his shattering defeat in Vienna. He was tailor-made for Paris, where wit, flair, conversational brilliance, and a high-stepping intellectual style provided a useful backdrop for his treatment center. As in Vienna, his "cures" were highly heralded by his well-to-do patients, for whom his clinic became a fashionable social center.

Mesmer asserted his willingness and indeed his eagerness to have his theories examined by competent bodies, but when a representative of the Royal Academy of Science in Paris volunteered to arrange for such an authoritative appraisal Mesmer demurred, giving the impression he thought his achievements were too far advanced to warrant such scrutiny. He did, however, submit to the Paris Medical Faculty a document summarizing his theories in twenty-seven propositions in which he restated his belief in the universal fluid, a specific expression of which was animal magnetism. He also put his emphasis on the need for the human body to maintain a condition of internal harmony and balance.

The scientific community in Paris became increasingly skeptical and disturbed, especially when it learned of the Societies of Harmony with their extravagant dues, to say nothing of the prospect that amateurs would use their newly discovered magnetic talents to treat themselves or their friends. The drumbeat of the professionals against Mesmer reached a pitch where not even Marie Antoinette or Lafayette could prevent the king's appointment of an investigating body.

Franklin's commission took its assignment seriously. With the concurrence of D'Eslon and Mesmer, the members attended the group treatment center with its mirrored walls, expensive tapestries, and thick carpets. They observed the famous Mesmerian *baquets*—oak buckets about two feet high, inside which were pebbles and bottles filled with "magnetized" water. From the apertures in the wooden covering over the tubs protruded L-shaped iron rods, which would be grasped by the patients sitting around the *baquets*.

The commission members watched the patients as they were led into a sort of collective hysteria, with Mesmer in his lavender silk robe waving his wand and orchestrating the movements of the group, getting them to sing or chant as they circled the *baquet* in a trancelike state, a rope connecting them to one another. The commissioners observed Mesmer and D'Eslon making "passes" at their patients, many of whom testified to their improvement or cure. The commissioners even submitted themselves to the magnetic procedures. Those of them who had ailments went through the full treatment.

Almost five months to the day after the investigation began, the commission came to a decision remarkably similar to the one reached by the medical faculty in Vienna. They found no evidence to support the theory of animal magnetism or a vital fluid. They tested the rods from the *baquets* and found no impulses, electrical or otherwise, emanating therefrom. They accepted the genuineness of some of the cures, which they attributed to changes produced by hysteria and the workings of the imagination. "Imagination without magnetism produces convulsions but magnetism without imagination produces nothing."

Franklin, however, was able to look beyond the showmanship and the charlatanry to the potential value of Mesmer's ideas. Franklin was fascinated with the evidence that powers of suggestion could produce physical changes. The commission's report did not emphasize this point but it did refer to the possibility of conversion of belief into bodily changes.

The general report was made public; but the commission sent a private report to the king.

Antoine-Laurent de Jussieu issued a separate report. He was aware of the false claims made by the Mesmerists, but he recognized that something significant happened during the exchange between practitioner and the patient. Whatever Mesmer's explanations, the fact remained that many of the cures were real. One did not dispose of these phenomena, he contended, just by reference to the imagination. For if, by invoking the imagination, disease or disorder could be reduced or dispatched, then a vast new resource in the treatment of disease had to be recognized.

As chairman, Franklin went along with the majority report, but his letters to friends reflected the view that one had to respect intangibles and imponderables in the treatment of disease. The power that resided in belief was manifestly true, but it was not necessary to repudiate medical science in order to comprehend forces exerted by the human mind.

Following the Franklin Commission report, Mesmer faded into the background, eventually returning to Vienna, where he fell into political disfavor with the nervous aristocrats because of his supposed sympathy with the French Revolutionists.

In 1812, at the age of seventy-eight, Mesmer finally achieved acceptance by medical faculties in Prussia and Berlin. In his eighty-first year, he died in a village not far from his birthplace. Cause of death was a malignancy of the bladder, untreatable at the time by any system, including his own.

Franz Anton Mesmer did not invent the term *hypnotism*. This particular word was introduced into the language in 1843 by Dr.

James Braid, a Scottish surgeon. Having seen "mesmerism" demonstrated, Braid developed it into his own version called "neurypnology," or "nervous sleep." He emphasized the subject's ability to convert thoughts into bodily change, thus enabling the physician to use powers of suggestion as a way of treating human illness.

Dr. Braid also drew upon the work of the Marquis de Puységur, a contemporary of Mesmer, in putting patients into a trance or sleep state (rather than Mesmer's therapeutic "fit") as an essential part of medical treatment. Braid was able, through hypnotic techniques, to relieve a great variety of ailments and even to perform major surgery without inflicting unbearable pain on his patients. A contemporary English surgeon, James Esdaile, soon reported hundreds of such cases.

The developing knowledge of hypnotism made a profound impression on Jean Martin Charcot, one of the most influential medical figures of the late nineteenth century. Charcot's studies led to his theories of "conversion hysteria"—the tendency of the mind, when gripped by forebodings or apprehensions, to produce illness. Charcot believed that Mesmer's "successes" were mainly with patients where, to begin with, illness may have involved a degree of conversion hysteria. Charcot was Freud's teacher and was able to imbue his student with a profound respect, not just for hypnotism but, more precisely, for the way the human mind was able to transfer its preoccupations and concerns into physiological reality.

The development of hypnotism in medicine has not come forward in a straight line since Mesmer. For a long time, it was regarded by a large segment of the medical profession as having little more than marginal value, except perhaps in psychiatry. But the increasing importance being assigned to psychological factors in causing serious disease has set the stage for the careful use of hypnosis as a valid part of medical treatment. When, for example, it seems clear to the physician that traditional measures are not producing the desired results, the physician may bring in a medical hypnotist—generally, a fully accredited physician who also holds credentials in hypnotism. In this context, hypnotism has been used to assist in childbearing and various dental procedures and to treat chronic pain syndromes, insomnia, alcoholism, cigarette addiction, drug addiction, arthritic symptoms, sight and hearing impairment, and many other disorders, including those that may have their origin in profound emotional stress. It has been used for certain types of psychiatric disorders, especially during wartime, for the past century. Hypnosis in some cases can also be a valuable aid in relieving panic and otherwise changing or controlling emotional states that may be involved in disease.

Medical researchers have produced evidence that prolonged depression or grief can have negative effects on the immune system, reducing the body's ability to combat hostile microorganisms or to cope with abnormal changes in cell growth. Similarly, panic is an intensifier of disease. One reason so many persons die of heart attacks during the first twenty-four hours is that the panic constricts the blood vessels, imposing an additional and sometimes intolerable burden on an already damaged heart. Also, panic can result in a sudden increase in certain hormones that can have the effect of further destabilizing the heart.

Unfortunately, the popular notion of hypnosis has been formed largely through vaudeville shows or their equivalents. Members of an audience are called up and are made to do things in response to the orders of a nonmedical hypnotist that are usually outside the range of normal behavior. Then the subjects will be returned to their senses, frequently to their own consternation and to the amusement of the audience. But the ability of the human mind to be steered in certain directions has opened realistic possibilities for bringing about biochemical changes inside the human body.

Modern research, as mentioned earlier, provides a picture of the human brain not just as the seat of consciousness but as a gland producing at least several dozen basic secretions that have been identified so far. Dr. Richard Bergland, a neurosurgeon affiliated with the Harvard Medical School, has pointed to the ability of the brain to combine these secretions. Not all these secretions are locked within the nervous system. What a person does or thinks can affect the kinds of prescriptions written for the body by the brain. Our ability to sustain severe pain can be enhanced as a result of substances produced in the brain. These chemicals have morphinelike molecules. The brain can also produce substances that can combat infections or that have a vital role in maintaining the body's essential balances.

Because of the tendency of extraordinary physical or emotional stress to upset these vital balances, anything that can reverse or block such stress may help to restore the body's normal recuperative functions. Medical hypnotists in many cases are able to help the patient to modify the offending emotions, thus freeing the body to assert its normalized functions.

Obviously, hypnosis can be dangerous. Jolly West, who has published a book on the history of hypnotism, has called attention to the horrors that have sometimes been produced for political or ideological purposes. For the hypnotist is able, in some cases, to eradicate or change memory, or to cause an individual to lie without any realization of the fact.

None of these developments, of course, is chargeable to Franz Anton Mesmer. It is significant, nonetheless, that the Franklin commission of 1784 called attention to both the ominous and some of the benevolent possibilities of mind control.

A modern example of mesmerism, or group hypnosis, may be observable in "firewalking," in which large numbers of people are "prepared" by powerful suggestions to endure the experience of walking over hot ashes.

13

FIREWALKING

Drs. Lee and Joyce Shulman, a married couple, are psychological consultants, with offices in Detroit and Los Angeles. They are also our old-time friends.

Knowing of my interest in the ability of the mind to produce biochemical change as the result of words or attitudes, Joyce telephoned about what she described as one of the most unusual experiences of her life. She had gone to a "firewalk"—an affair in which more than a hundred persons in bare feet had walked across a bed of hot coals. She had gone out of curiosity, became caught up in the fervor that preceded the exercise, took off her shoes, and joined the procession of chanting, wide-eyed people as they demonstrated their ability to subject their feet to a fiery experience without being burned. Joyce came through unharmed.

"It's something you really ought to see for yourself," she said. "I told Lee about it. He is fascinated and plans to go to the next one.

"What happens is that a young man named Tony Robbins prepares the people for the experience emotionally," she continued. "It's a group session that takes about two hours. He gets the people to believe that they have the power to expose their feet to fiery heat without injury. And they do it."

I didn't respond at once, trying to take in the significance of what she was saying.

"Do you hear me?" she asked.

"Yes, I hear you. How do you explain it?"

"Lee and I have discussed it. He's skeptical, of course. He would have to be. He wasn't there. I was and it's real."

Firewalking, of course, is nothing new. Tribal rituals have used such exercises to propitiate the gods or to demonstrate priestly powers. The press had run stories about firewalks, but they seemed to be in the category of odd events. I told Joyce that I shared Lee's skepticism.

"In that case, you might want to observe it for yourself. Tony Robbins will be conducting another firewalk in a couple of weeks. Lee and I will be going. Why don't you and Ellen join us?"

Thus it was that Ellen and I, in company with the Shulmans, had an opportunity to attend one of the most unusual and certainly one of the strangest group demonstrations I had ever witnessed. It took place in a private auditorium in the heart of Los Angeles. About 150 or 200 people sat in rapt attention while Tony Robbins, a tall and charismatic young man, led them through a session that at times seemed like a serious lecture on human potentiality; at times like a Werner Erhard est session in which the audience is cajoled, prodded, praised, pummeled; at times like a Dale Carnegie session on the secrets of success; at times like a Holy Roller revival meeting with people singing, chanting, reaching out, embracing one another.

If you threw yourself into the spirit of the occasion, you discovered that you had delegated a large part of your critical judgment to Tony Robbins. The man had extraordinary stage presence. He was like a music conductor, summoning some of the instruments to soothe or sedate, or invoking the entire orchestra in a pounding crescendo. Finally, after creating a single entity out of an aggregation of varying individuals, he told them they had the power to walk barefoot over hot coals and be completely impervious to the heat.

"It will feel like cool moss," he said. "We will all chant 'Cool moss' as we file downstairs."

I looked around. The people were transported. It was an exercise in mass hypnotism. This was the way, I thought, the people in Mesmer's Paris salon must have looked, as, chanting and trancelike, they circled around the wooden tub in the "animal magnetism" ritual.

Back of the auditorium, in a small yard, workmen were piling logs on a firebox about eight or nine feet long and a little more than three feet wide. The fire raged upward, sending out fierce waves of heat. Then, after the fire subsided and while the ashes were still alive, the people, pants rolled up or dresses raised, lined up for the culminating experience. Tony Robbins stood at the head of the firebox, giving last-minute instructions, his hands on each person before releasing them onto the hot coals. Everyone was still chanting "Cool moss, cool moss." The chant was raised to a cry by most of the people as they hurried across the hot ashes. Some of them, on completing the journey, would sing out in triumph. Several of them were openly hysterical and wept on coming through.

One of Tony Robbins's assistants came up to me and said that Robbins was eager to know whether I would like to go across the firebox. I thanked the gentleman and declined.

To my astonishment, I saw that Lee Shulman was on line, standing behind Joyce, ready to take the walk. Joyce, smiling broadly, took four large steps across the firebox, then looked back triumphantly and encouragingly at Lee. Tony Robbins had his hand on Lee's shoulders and was saying something in his ear. Then, chanting "Cool moss, cool moss," Lee started out across the coals. At the far end, his face became contorted in pain. Joyce, concern on her face, started to attend to him. He had been burned. I rushed to his side.

"It's not bad," he said. "I got some hot cinders between my toes. They're out now."

Was any significance to be attached to the fact that Joyce, a true believer, had been impervious to the heat while her husband, curious but skeptical, had been slightly burned? Or were the cinders between his toes a consequence of the fact that he stomped through the firebox, penetrating the soft and insulating ash?

There were other questions—pertaining not to Lee but to the entire affair. I had noticed that the workmen had wetted down the ground at the beginning and the far end of the firebox. Could the moisture on the soles of the feet have provided some insulation, at least, for the very short journey? Next, what about the "hot coals"? The firewalkers were not stepping on hot coals but on a thick insulating layer of soft ash. Didn't that make a difference? Why was the firebox so short—requiring only four steps for most of the walkers? Suppose the box had been twice or three times the length?

Yet—and here we shift to the other side—did the belief system have no role at all in outcome? Lee Shulman, as mentioned a moment ago, carried a large burden of rational doubt with him when he started across the firebox. Joyce, on the other hand, felt completely transported.

Was it possible that the belief system provided limited protection within a certain range—and that the range varied with the individual? Did the preparation session produce an element both of mass and self-hypnosis that conferred a margin of resistance to the heat? Did the people who trod more heavily than others have increased risk because their feet penetrated through to the hot cinders?

I left the firewalk experience with more questions than when I came. The emotional preparation session inside the hall had had a profound effect on most of the people, as one could plainly see by looking at their eyes and listening to their chants. How essential was the emotional workup of that session to the result? If there had been no such preparation, would as many people have been able to sustain the experience? More pertinent still: Would they have had the desire to do so?

At a roundtable meeting of psychiatrists, I recounted the firewalk episode. Almost every person present had a similar story to tell involving seemingly superhuman experiences. But the consensus of the group was that there is a point at which the powers of the mind cease to have control over physical forces. The group agreed that while mental powers (with or without hypnosis) could reduce the pain inflicted by extreme heat, those powers could not confer immunity to skin exposed to direct fire for more than a second or so. Anyone can run a finger through the flame of a candle, but the result would be entirely different if the finger were motionless inside the flame.

We were left with a series of relative hypotheses about the firewalk. Yes, the brain could help control pain. Dr. John C. Liebeskind, UCLA psychologist, was able to show that the body, under heightened emotional circumstances, could produce its own painkilling secretions. During the past ten years, I have witnessed various demonstrations of pain control. In company with a dozen or so members of the medical faculty, I have seen a man run a darning needle through his arm without drawing blood and without discernible pain. We had also seen another man lie down on a board through which protruded several dozen sharp nails. The nails penetrated his back and shoulders but there was no bleeding—except for one site on his right shoulder. When we called this to his attention, he thanked us and turned it off. But this control, impressive though it was, was still not absolute. It didn't mean that he could withstand an ax coming down on his arm, or a razor opening up a vein. Again, what we were really considering was the *range* within which the belief system could help affect bodily events, as well as the *extent* to which the conditioning process figures in such changes.

It is well known that people, even if they are not under trancelike conditions, can be told, while blindfolded, that a hot knife will be applied to the skin for only a second. A cold knife touches the skin very lightly and produces a burn blister. The heat, obviously, is furnished by the mind.

In assessing the significance of the firewalk, the psychiatrists at the roundtable hesitated to regard the experiment as providing verifiable evidence of the powers of mind. But they had no doubt that the belief system *did* have some effect—how much, no one in the group could say. Yet even a small effect was interesting and significant. It indicated the existence of pathways along which the belief system could travel.

The big need, therefore, was to study these pathways for clues to wider application. Even a small clue that attitudes could produce biochemical or biological change could be valuable in indicating a research direction that might lead to highly significant findings. In any case, the firewalk phenomenon was a bead, albeit a small one, that was available for fitting onto our string.

Meanwhile, the enormous literature on hypnotic anesthesia, from Braid and Esdaile in the 1840s to the present day, remains before us. According to Jolly West, this literature should probably also include the phenomenology of acupuncture, which he believes works in the same way that hypnosis does—as an anesthetic.

14

THE PHYSICIAN IN LITERATURE

The first assignment given me at the medical school was to teach a course on "The Physician in Literature." This provided a good opportunity to show the connection between the emotions and illness, since writers find this a favorite theme.

For example, Boris Pasternak, in *Dr. Zhivago*, describes a conversation between Zhivago and an old friend, Mischa. Zhivago reveals that he has been having some cardiac difficulties but he looks beyond physical factors for the answer.

"It's a typical modern disease," Zhivago says. "I think its causes are of a moral order. The great majority of us are required to live a life of constant systematic duplicity. Your health is bound to be affected if, day after day, you say the opposite of what you feel, if you grovel before what you dislike and rejoice at what brings you nothing but misfortune. Our nervous system isn't just a fiction; it's a part of our physical body, and our soul exists in space and is inside us, like the teeth in our mouth. It can't be forever violated with impunity."

In that one paragraph, Pasternak not only offers a compressed history of Russian postrevolutionary years but anticipates what is perhaps the most important development in modern medicine—the recognition of a biochemistry of the emotions, underwritten by fast-accumulating evidence of specific interactions between the brain and the endocrine and immune systems. Many contemporary medical researchers believe that bacteriological, viral, and other organisms are a secondary and not a primary cause of illness. Before these organisms can take hold they must first crack the barrier of the human immune system—a spectacular assemblage of sentries and defenders with highly developed capabilities for poisoning or crushing intruders. Only when the immune system is impaired or depleted are disease agents able to penetrate the defenses. The evidence is incontrovertible, as Pasternak predicted, that emotional wear and tear figure in that impairment or depletion. Here a vicious circle can be discerned. Serious illness almost universally produces depression; the affected individual feels loss of control and seems helpless to stop onrushing events. Those emotions work their way from the brain into hormonal and immune functions, intensifying the underlying problem with consequent deepening of helplessness and depression. Susan Sontag's *Illness as Metaphor* is an interesting postscript to Pasternak's perception of what was happening to the Russian soul in the long night of political depersonalization.

Dr. Zhivago was an effective doorway between medicine and literature in my lectures to medical students. The course I gave at UCLA was on "The Physician in Literature," the title of a book I had published only a year or two earlier. Dean Mellinkoff told me that an unusually large number of students had signed up for the course, augmented by about seventy-five persons from the surrounding community who asked permission to attend the lectures.

In connecting medicine to the humanities Dean Mellinkoff hoped to do something about the fact that incoming students were long on science and short on the arts. This was understandable; the liberal arts had been downplayed in medical education. The assumption was made that undergraduate education met the rounded needs of a student and that, therefore, the graduate school could afford to concentrate on the particularized requirements that go with a professional career. But the assumption was unjustified. Undergraduate students who intended to go on to medical school tended to steer away from the liberal arts. As mentioned in an earlier chapter, they did so in the belief that their chances for admission to medical school would be strengthened in direct proportion to their demonstrated excellence in the sciences. The result was that they put most of their scholastic energies into studies that dealt with quantifiable matters—and arrived at medical school in a state of educational disequilibrium.

This was not a minor deprivation. It affected their total ability to deal with the complex equation represented by illness. Such illness is often the result not just of an encounter with a pathological organism but of a way of life. The physician, therefore, must be able to assess the full range of causes involved in the patient's illness. Anyone who goes into medicine must understand the ease with which modern society transfers its malaises to the individual. He must comprehend the variations of stress in modern family life and in human relationships in general. He must understand disappointment, rejection, blocked exits. He must have an appreciation of what is required to make the individual whole again. Literature helps the medical student to make connections between the experiences of the race and the condition of the individual, and to fit the individual into a world that is not as congenial as it ought to be for people who are more fragile than they need to be.

Almost every literary figure of any consequence—from Aeschylus to Walker Percy—has had something to say about doctors. To the writer, the physician is not just a prescriber of medicaments but a symbol of all that is transferable from one human to another short of immortality. We may not be able to live forever, but most people persist in the notion that the physician possesses the science and the artistry that will provide them with endless deferrals. To be able to listen to the human heart and draw meaning from its slightest vibrations or whispers, to be able to take a tiny droplet of blood and perceive its vital balances, to convert electric markings into precise knowledge of the body's chemical complexities—all these may represent science to the doctor, but to the patient they are powers that come from the gods.

Writers have not had to imagine the patient's condition. They know at first hand all the frailties and uncertainties and loneliness that the physician is expected to banish. For writers, all too often, have been deeply troubled patients themselves. They have a direct acquaintance with the doctor's little black bag and know the value of the physician's touch and the healing power of his presence.

Imagination and the art of writing go together. And, since imagination is the basic ingredient of hypochondria, writers have had little difficulty in suffering from all sorts of symptoms. Consider Proust, his windows shut and his blinds tightly drawn, seldom emerging from his flat, living in a habitat certain to intensify his ailments, only a few of which were organic but all of which kindled his imagination. The greater the writer's uncertainty about his health, the greater the dependence on the physician.

It is here that we observe an element of reverence in the attitude of many writers toward physicians—if not of their persons, then of their roles. For when the writer becomes a patient, his dependence on the physician is no less epic than it is with anyone else. Naturally, this leads to unreasonable expectations of the powers of the physician. Consider as evidence the way many writers describe the arrival of physicians at a critical time. The family is huddled in dreadful uncertainty around the bedside; but the presence of the physician produces a miraculous change of spirits. It is not always possible, of course, for the physician to work magic; reality provides a good test of the writer's ability to deal with heightened emotions.

It is natural that writers, in dealing with ultimate confrontations, should find the doctor such a convenient literary resource. Physicians may vary widely in their personal and philosophical behavior, but they all offer rich material to the novelist. Voltaire writes dispassionately in *Candide* of the helplessness of physicians when confronted by venereal disease—a penalty attached to the deepest of pleasures. Alexander Pope takes such paradoxes in stride but also holds the view that the physician should not be expected to be kinder than God. Boris Pasternak, true to the Russian literary tradition, sees the physician as the embodiment of all the mixed emotions that the human tragicomedy can produce. There is no standardization of attitudes or responses—nor can there be—but the physician never lets the writer down. The entire field of medicine offers infinite materials to the novelist.

Of all the Russian writers, perhaps none exhibited greater fascination with the doctor than Tolstoy. He brings all his gifts of descriptive irony to the account of the physician who treats Ivan Ilyich (*The Death of Ivan Ilyich*). The doctor is supreme, invulnerable, possesses a confident claim on the future. The patient is dependent, tentative, a sure loser. A somewhat similar perception of the physician emerges in Dostoevski's *Crime and Punishment*. The sufferings of the ill are far more memorably described than are manifestations of compassion by the physician. Even when the physician is in love with his patient, as in Turgenev's "The District Doctor," there is a perceptible distance in the relationship between the doctor and the object of his affections. Is this a reflection of the traditional reserve doctors are supposed to maintain in their relationship with patients? (Medical students will think of *Aequanimitas*, by Sir William Osler.) Or is it a manifestation of the eternal loneliness in the Russian soul, portrayed so powerfully by Gogol in "The Diary of a Madman"? What the Russian writers are trying to tell us, perhaps, is that human beings are never really able to shatter their loneliness and that physicians, whatever their magic, are capable only of limited rescues. This generality is not as morbid as it seems; there is always the redeeming virtue of an interim triumph, the discovery that improvement and prolongation of life are attainable prizes and therefore can also be a sustaining reality.

In the English novel, the physician tends to be treated more as an institution than as a person. One thinks of Emily Brontë's description in *Wuthering Heights* of the physician whose professional detachment keeps him from becoming a vitally needed emotional resource for the patient. Not that the physican is callous; he is acting in the institutionalized way that doctors are supposed to act. The physician whose towering sense of authority leads him to make arbitrary decisions is seldom better described than in Wilkie Collins's *The Moonstone*. Thomas Hardy's Dr. Fitzpiers in *The Woodlanders* seems to feel his special station entitles him to exploit the affections of the heroine. Samuel Butler's observations about the institution of the doctor may be no more satirical than those of his contemporaries, but there is no mistaking the underlying resentments that are sometimes part of the physician's quixotic attitudes toward his patients.

What about the physician's vaunted heroic role? In all cultures, the role is as real as it is substantial. If the demigod concept is resisted by a minority of novelists, it is not because writers are unmindful of the exalted station of the doctor, but because novelists are merchants of paradox and searchers after warts. Even when they are unabashedly idealistic, writers reveal in their work their conviction that reality is best portrayed through contradictions. And the physician's calling enables him to preside over lives in a way that is the envy of politicians and, indeed, of all those who by calling or temperament try to steer people. It is inevitable that physicians, who are supposed to possess life-and-death powers, should be so often idealized by most people and ascribed with virtues equal to their authority. But the novelist is careful not to ignore the juxtapositions and complexities of character that make the physician credible.

In any case, the writer's fascination with the physician has been a continuing characteristic of world literature—and the American novel is no exception. Hemingway, both in his short stories and novels, makes use of his special knowledge as a doctor's son. The loneliness and despair of the ill people in his books are in stark contrast to the lofty estate of the physician. For the physician comes and goes at will, in contrast to the desperation and immobility of the patient. William Faulkner's doctors are perhaps more philosphical than are Hemingway's; they look not just at illness but at all life. In the same vein, John Steinbeck's Doc Burton, from *In Dubious Battle*, manifests a curiosity not just about the workings of the human body but about society; he is thereby justified in addressing himself to the interaction between the two. Even when Steinbeck's doctors are erratic and deeply flawed, as in *Cannery Row*, they remain the rallying points for a surrounding humanity. Walker Percy, himself a doctor, sees tragic human weaknesses in doctors, especially in his *Love in the Ruins*, but he also sees a vital human spark inherent in the physician that touches off life-giving energies and prospects. Ring Lardner, chronicler of picturesque character and speech, portrays the physician not just as a mediator with death but as a protector of those who are easily gulled or made to look ridiculous in the eyes of their fellows.

From literature, we learn that the responsible physician is more than just a scientist able to make a difficult diagnosis; he is a human being whose skill depends as much on his knowledge of life as it does on his knowledge of disease. Proper treatment calls for an awareness of human uniqueness and for sensitivity to all the elements of human potentiality. Poetry cannot replace prescriptions but it can widen perceptions. The best education for the physician is a blend of science and the liberal arts. A knowledge of thc clusive aspects of human uniqueness is no less important than technical aids used by the physician in combating disease.

What the world's great literature tells us about medicine is that few things are more important than the psychological management of the patient. Hippocrates and Galen and the other early greats of medicine may not have known about endorphins, enkephalins, gamma globulins, epinephrine, interferon, and the entire range of messengers from the brain to the body's organs. But they knew a great deal about the totality of the human organism and the interaction of all its parts. A person's outlook on life can be a vital factor in the onset and course of a disease. The wise physician, when making a prognosis, does not confine himself or herself to the virulence of the particular microorganism involved or the nature of an abnormal growth; the wise physician makes a careful estimate of the patient's will to live and the ability to put to work all the resources of spirit that can be translated into beneficial biochemical changes.

Writers such as Dickens, Hardy, Tolstoy, and Dostoevski subjected their various characters to recurrent strain. They would serialize their stories prior to publication in book form; each installment would end with an impending crisis in order to arouse interest in subsequent chapters. The extraordinary trials endured by the fictional characters would have produced any number of cases of emotional exhaustion in real life. Obviously, we don't need great novels to tell us that the ability of human beings to tolerate stress is incontestably finite. But those same novels help us to recognize that attitudes are vital factors in enabling people to meet serious problems, whether they take the form of illness or crises of circumstance.

Oliver Wendell Holmes, one of America's most distinguished physicians and philosophers of medicine, once proposed some perennial questions for doctors:

How does your knowledge stand today?
What must you expect to forget?
What remains for you to learn?

Winds of change now blow throughout American medicine, and one of the most promising zephyrs is the growing recognition that a good medical education involves more than science. The questions Dr. Holmes proposed are essentially philosophical. They cannot be answered without reference to intellectual and scientific history; nor without relating education and occupation to the needs of the society; nor without retrospective and prospective compass points. They cannot be adequately answered without some exposure to the cluster of intellectual disciplines that come under the heading of the humanities, which embrace not just the general range of human experience but the creative arts and the way people come to terms with life.

Science puts its emphasis on research and verifiable fact. Art and philosophy put the emphasis on the creative process and on values—values that come out of the memory of the race and that have something to do with the importance of being human, values that are conscious respecters of the unknown factors in the human equation. Among the recent discoveries in the practice of medicine is the fact that human beings come equipped with resources for healing that are best mobilized not by detached scientific efficiency, but by communication and supportive human outreach.

Basic to any education is one unchanging fact—that is, that facts do not stand still. A great deal of what medical students are now learning in their formal scientific education will become outdated within a decade or two after graduation. It is obviously and remorselessly true that the factual base of medicine has steadily changed in response to new findings about the nature of disease and the treatment of disease.

What endures, too, is the system for teaching scientific knowledge even if the knowledge itself tends to be fragile. I refer to the scientific method. The *way* new facts are discovered and developed; the *way* these facts are scrutinized and put to the test; in short, the *way* theory is translated into practice—this is what endures and what gives science its essential character.

Respect for the scientific method is a vital ingredient in any medical education.

There is no conflict between the scientific method and the need in the medical curriculum for subjects that deal with human values. Values constitute a moral system that transcends change. When values are strong enough and good enough, changes in science can be fitted into the lives of people, making it unnecessary to fit people into change. The way people are treated as patients can be as important as the physical things that are done in the attempt to ease or cure their ills. The effectiveness of the doctor as scientist is tied to his or her qualifications as artist and philosopher—intangible credentials that have to do with character and personal dimensions.

The auspicious news is that many medical schools, including UCLA, have recently broadened their admission policies to reflect respect for the liberal arts. The separate paths that the sciences and humanities have taken in search of truth are now converging in the wake of new findings. Human survival may depend upon the ability of human beings to work within nature rather than in opposition to it—as well as upon the ability to control the proliferation of knowledge that threatens to overwhelm us. The convergence is bringing about a new unity that cuts across disciplines. We are seeing a new breed of scientific humanists and humanistic scientists. The separation of the two intellectual worlds is giving way to a realization that they are both dependent on the conditions of creativity in the modern world.

The trend has been moving away from scientists who make public proclamations about the morally antiseptic nature of their calling, who detach themselves from the effects of their theories and discoveries. By contrast, more and more scientists insist that they are in a better position to understand the significance and implications for society of their discoveries than are the official decision-makers who may be paying their salaries or subsidizing their work. And, just as scientists are divided, so humanists are split on issues of human values. The point is that the real division is no longer between science and the humanities—the two cultures described by scientist-philosopher C. P. Snow—but between those who attach primary importance to human life and those who view their own discipline as sovereign.

The explosive proliferation of scientific knowledge in the past few decades has left knowledgeable members of the human species feeling unsettled, uncertain, even out of control. Young people have good reason to question the adequacy of an education that has separated them from the questions that bear upon their own future, the future of mankind, and the quality of life—all of which have a bearing on health.

Common to the sciences and the humanities is the human urge to understand the universe and man's connection to it. The failures that have pockmarked history have come at times of philosophical poverty. Humans may enlarge their objective techniques and even their knowledge, but they cannot change the basic fact that their position in contemplating the great questions is inherently subjective.

The science and art of medicine converge at the point where physicians become basically concerned—as, traditionally, poets have been—with the whole of the human condition. "I feel convinced," wrote Claude Bernard, "that there will come a day when physiologists, poets, and philosophers will all speak the same language."

15

TASK FORCE BEGINNINGS

My dream, after being at UCLA a few years, was that the medical school might institute a program or department in mind-body studies—a program that would employ the scientific method to explore ways in which attitudes or emotions are converted into biochemical change.

The dream began to be converted into reality when I was informed by Dean Mellinkoff in July 1984 that Mrs. Joan Kroc of San Diego hoped that we might have lunch. Mrs. Kroc was the widow of Ray Kroc, the founder and owner of the McDonald's chain of fast-food restaurants. A date was arranged for San Diego.

We met at the La Valencia Hotel in La Jolla. Mrs. Kroc was accompanied by her lawyer, Mrs. Elizabeth Benes. Within a few minutes after sitting down to lunch, I had a sense of being in the presence of two persons who were extraordinarily well informed about national and world affairs and about educational and social problems. I learned that Mrs. Kroc had created a foundation, which she and Mrs. Benes directed and operated. The general concern of the foundation was the human condition, and it was especially active in combating alcohol and drug abuse.

After describing the activities of the foundation, Mrs. Kroc said she was familiar in a general way with my work and that she would like to endow a permanent chair in my name at UCLA. Mrs. Benes said that her inquiries indicated that $2 million might be adequate for that purpose. She asked whether I would be agreeable to the gift.

I almost fell off my seat. One thing I do know was that I stammered to the point of incoherence. When I regained my equilibrium I tried to express my profound gratitude and asked whether the main purpose of the gift was to support a certain line of work or an individual. Mrs. Kroc said both. I said that endowing a chair was a great personal honor but I would feel awkward, as a new faculty member, having my name on it. I wondered whether she would consider making it a "purpose" gift; that is, whether it could be given for the specific purpose of advancing research in mind-body studies.

Mrs. Kroc and Mrs. Benes readily agreed. They also agreed to my proposal that a portion of the gift be set aside to support similiar research at other medical centers. UCLA would make recommendations to the Joan B. Kroc Foundation about such grants, the purpose of which would enable us to do networking in the field of psychoneuroimmunology, the term describing the interaction between the brain, the endocrine system, and the immune system.

When I returned to Los Angeles and reported to Dean Mellinkoff on the meeting in La Jolla, I learned that the gift from Mrs. Kroc would be one of the largest in the recent history of the medical school connected to specific research rather than a building or facilities.

Now came the exciting business of planning. In addition to seeking the advice of the dean, I had extensive discussions with Dr. West, Dr. Franklin Murphy, and Dr. Carmine Clemente. The consensus of these sessions was that we should create a functioning committee or task force that would serve as a sort of think tank, identifying opportunities for research that might be undertaken at UCLA and keeping track of work being done elsewhere that might warrant our support.

The formation of the task force was the biggest single step in the direction of our goal. It enabled us to have a specific mechanism with scientific credentials for pursuing the nature of the biochemistry of attitudes and emotions. Dr. Mellinkoff,

Dr. West, and Dr. Clemente agreed to constitute an advisory committee. Since we would be concerned with the interactions of the nervous system, the endocrine system, and the immune system, we would want the task force to be represented in these particular areas. In addition, we would be considering patient-physician relationships, and the general environment of medical care. We would want to be represented in these fields as well.

We were fortunate in being able to recruit all our first choices:

- *Dr. Claus B. Bahnson*, a Danish-born psychologist who, in the fifties, had begun to investigate the connection between emotional factors and cancer. Dr. Bahnson joined the Fresno medical facility of the University of California at San Francisco. The affiliation made him eligible for membership on our task force.
- *Dr. John L. Fahey*, probably the outstanding immunologist on the UCLA medical faculty and consultant to government medical agencies concerned with basic research on immune functions and interactions.
- *Dr. John C. Liebeskind*, UCLA psychologist, and a world-renowned scientist on the biology of pain and pain inhibition, whose laboratory established a role for the body's endogenous opiates in the immune system. He is the founding president of the International Pain Foundation.
- *Dr. George F. Solomon*, who had dual affiliation with the medical schools of UCLA and the University of California at San Francisco. One of the recognized pioneers in the emerging field of psychoneuroimmunology, he was a contributor to the landmark book *Psychoneuroimmunology*, edited by Dr. Robert Ader of the University of Rochester. His basic training was in psychiatry but his clinical and laboratory work spanned a wide range of medical problems, laying the foundation for his psychoneuroimmunologic research.
- *Dr. Bernard Towers*, English anatomist-philosopher and medical ethicist who had joined the UCLA medical faculty some years earlier and who had established a program in medicine, law, ethics, and human values. He had also originated a forum series for public, faculty, and students on vital medical issues. More recently, he has worked directly with seriously ill patients,

using imagery as a means of developing their own resources, in combination with medical treatment, for combating illness.

This original group was later augmented by:

- *Dr. Fawzy I. Fawzy*, UCLA psychiatrist, who had considerable experience in working with cancer and AIDS patients, and who was the medical chief of a research project at UCLA involving a substantial number of malignant melanoma patients in an effort to see whether positive attitudes could affect the course of disease.
- *Dr. Herbert Weiner*, UCLA psychiatrist, who had done early studies on the connection between behavioral factors and serious disease. His work *Psychobiology*, published in 1976, was probably the most advanced and complete account of the field at that time.
- *Dr. Carmine Clemente*, whose retirement as head of the UCLA Brain Research Institute made it possible for him to move from an advisory to an active role on the task force.

The first meeting of our new Task Force in Psychoneuroimmunology took place in the Jacoby Conference Room of the Jonsson Comprehensive Cancer Center, the building in which my own office was located. Mrs. Kroc, at my urging, attended that first meeting and expressed her appreciation to the group for its willingness to monitor the emerging field of psychoneuroimmunology and to identify opportunities for useful research. She said she was deeply interested in the objective of our work, feeling that it had useful implications for health and medical treatment.

After thanking Mrs. Kroc, I outlined a proposed plan of action for the group's approval. Her gift of $2 million would be divided into two equal parts—one part for research at UCLA and the other to help support similar research at medical centers elsewhere. Our aim was to study the biochemistry of emotional states. How did hope, the will to live, faith, laughter, festivity, purpose, determination register in the brain? What effect did they have on the body? Was it possible that the brain might play an active part in the healing process? Might the brain be consciously directed for that purpose?

Dr. Towers said the group was being asked, in effect, to provide a physiological explanation for human uniqueness. Wasn't that a tall order?

I said I could think of no more useful or exciting assignment. In any case, we were trying to create an arena for scrutinizing the wide range of factors—and not merely the mechanistic ones—involved in the treatment of disease. The view of the brain as the seat of consciousness was being extended to the concept of the brain as a biological control post and gland regulator that both sends and receives messages from other parts of the body. This shift furnished rich materials for our discussions.

Pursuing the practical implications of such research, Dr. Solomon described the structural and biochemical similarities between the brain and immune cells. If the approach to treating cancer involves the immune system, the brain should manifest alterations as well. In support of this hypothesis, it was pointed out that scientists are now finding that the administration of interferon (an immune regulator that also inhibits viral cell growth) promotes severe depression in AIDS patients.

Dr. Weiner noted that Dr. Liebeskind's work on endorphins had wide implications for these issues. Since morphine depresses natural killer cell function and promotes tumor growth, what happens when human pain is "managed" by opiates and other drugs? Dr. Liebeskind replied by saying he had been studying the effect of three different analgesics—morphine, fentanyl, and sufentanil (used in cardiac surgery as adjunct or sole anesthetics)—on NK cells. The three analgesics have a more or less equivalent inhibiting effect on NK cells. However, the more powerful of these drugs, such as fentanyl and sufentanil, have an even greater effect.

His comment pointed to one of the most fascinating and difficult questions confronting physicians: how to determine whether the benefits of a certain drug—or even a procedure—outweigh the possible risks. If a patient is experiencing severe pain, should the need to provide relief take precedence over the negative effects of the painkiller on the body's own system for combating the illness? A possible guideline is the extent to which a drug may impair the ability of immune cells to play a major role in combating a serious illness. In some cases, the administration of a drug may be moderated in order to achieve a combined effect. John Liebeskind pointed to the need for extensive research that could throw additional light on these questions.

Dr. Bahnson called attention to the danger of viewing the body as a set of compartmentalized functions in the treatment of disease. We had to recognize that drugs affected not just the parts but the whole. Dr. Weiner concurred, emphasizing that traditional ways of splitting up bodily systems are no longer acceptable. For example, why should there be neurotransmitter receptors on lymphocytes in the spleen? Why should they enable lymphocytes to be stimulated by the nervous sytem, as in Dr. David Felten's work at the University of Rochester? We have to outgrow the notion of separate systems. Everything has an effect on everything else.

Dr. Clemente noted that the field of biomedicine has become more atomic and molecular. He recognized that the questions we were posing had to be addressed—and the Task Force was in an excellent position to undertake these probes. The evidence was growing, he said, that the brain was a connecting link among the endocrine, immune, and autonomic nervous systems. He doubted it was any longer correct to talk about a wall of separation blocking the conscious intelligence from such functions as cell reproduction, or hormonal activity, or circulation of the blood. He saw no reason why these questions shouldn't be investigated. If, in fact, specific knowledge could be developed in these directions, then the gain not just to medicine but to society could be beyond calculation. Dr. Clemente said that the resources of the Brain Research Institute were at the disposal of the group.

Dr. Fahey said that a great deal of thought had to go into the protocol for such research. He wondered whether even the new resources available to us might be sufficient for what was contemplated. A great deal of work had to be done if we wanted to show that positive as well as negative attitudes had an effect on the immune system. Nonetheless, in general agreement with Dr. Clemente, Dr. Fahey said we ought to get started. He reminded the group that several years ago, Dr. Liebeskind had said that they should go ahead with information at hand in investigating the analgesic response to pain that led to the discovery of the role of endorphins and enkephalins in the suppression of pain. This research in turn led to the study of the way such natural analgesics also activated the immune system. According to Dr. Fahey, there was now abundant evidence of real dynamics involved in psychoneuroimmunologic research.

The value of our discussions, I suggested, was that we were really creating a serendipitous arena. Any good laboratory or research center was a place where unexpected things could happen. A specific goal might be defined for a research effort, but just in the process of pursuing that goal, other possibilities might come about indirectly.

Hans Selye, in his *From Dream to Discovery*, had defined the essence of scientific discovery. "It is not to see something first," he said, "but to establish solid connections between the previously known and the hitherto unknown. . . . Some of the most fundamental discoveries in medicine have been and still are being made by people who use no complex machinery but only their intuitive feeling for the way Nature works and a keen eye for what she camouflages."

Along the same lines, we could consider this observation from Hans Zinsser's *As I Remember Him*: "Even Archimedes' sudden inspiration in the bathtub; Newton's experience in the apple orchard; Descartes' geometrical discoveries in his bed; Darwin's flash of ludicity on reading a passage in Malthus; . . . were not messages out of the blue. They were the final coordinations, by minds of genius, of innumerable accumulated facts and impressions which lesser men could grasp only in their uncorrelated isolation, but which—by them—were seen in entirety and integrated into general principles. . . . The one who places the last stone and steps across to the *terra firma* of accomplished discovery gets all the credit. Only the initiated know and honor those whose patient integrity and devotion to exact observation have made the last step possible."

Dr. Clemente recalled that each semester, when his first-year medical students dissect the spleen, they see nerves the size of the optic nerve—more than what seems necessary to stimulate the few blood vessels that are present. Since the spleen and the thymus are involved in immune function, the role of those nerves in the process could be assumed. In any case, we perceived a connection between these speculations and the work of the Task Force.

Dr. Mellinkoff, who came to some of our meetings, reminded the group that some of the greatest discoveries have been anticipated by teleological reasoning. He quoted Bichat who said of the liver, "It is not conceivable to me that nature made an organ so large for the sole purpose of excreting a liquid less copious than urine."

The consensus of the Task Force was that, so long as we were not expected to come up with *definitive* answers to questions about the physiology or biochemistry of the positive emotions but to think about the ways of studying the questions, and so long as we would be content with the pursuit and the possibility of serendipitous findings, we could probably justify our existence.

And so it went, each member drawing upon his primary medical interests. We recognized the need to keep ourselves up to date in mind-body research wherever it was undertaken, to discuss its implications, to map out research projects of our own, and to consider ways of furthering education and research in psychoneuroimmunology at UCLA and elsewhere.

We also recognized that we would need a "facilitator" who would prepare the agenda for each meeting, keep the minutes, prepare the reports of research projects conducted under our auspices, whether at UCLA or elsewhere. We had the extraordinary good fortune to find and hire Ms. Ping Ho, who had had experience in working with health research groups as rapporteur and historian.

After the meeting, Mrs. Kroc told me she was heartened by the approach of the group and deeply impressed with the stature of its members. As for me, at that particular stage, I couldn't ask for anything more. The obsession that brought me to UCLA was moving toward its goal; we now had a group of scientists to search systematically for answers to the basic questions connected to that obsession, and an open track on which to proceed.

At subsequent meetings of the Task Force, we considered opportunities for research. Dr. Weiner said we had to contend with the fact that such research might run counter to the traditional biomedical focus on pathological processes rather than on the role that bodily systems play on health maintenance. Dr. Solomon described another contributing factor to the problem of biased information: physicians rarely report unexplained remissions or recoveries. He informed the group that Brendan O'Regan of the Institute of Noetic Sciences in Sausalito, California, has compiled from medical journals more than three thousand reports of spontaneous remissions in immune-related disorders. ("Noetic" refers to the role of the mind in human affairs.) Although there has been a great deal of variability in the degree of information provided by each account, both psychosocial and immunological factors appeared to be involved in remissions.

A fuller understanding of health and illness, Dr. Solomon said, made it necessary to examine all the factors at work in the intensification or retreat of illness. Stress, for example, could increase illness; eliminating the stress could aid the recovery process. He emphasized the importance of studying the contribution of optimal emotional *health* toward physical well-being.

Weren't medical studies more concerned with breakdown than with recovery? I asked. A great deal is taught about pathology; how much was being taught about healing? Earlier, I mentioned the *New England Journal of Medicine* article by Franz Ingelfinger which reported that 85 percent of all ailments are self-limiting. What, specifically, was known about the process? How does the body itself manufacture healing substances adequate to its needs? How does it know when to start the repair process, when to speed it up, when to stop? The work of the Task Force should aim at restoring a balance between medical science and knowledge of the resources built into the human body. Was it likely that the next development in medicine would flow out of increased knowledge of the nature and workings of those resources?

At one of our meetings, Dr. Bahnson raised the issue of the patient-physician relationship. Both he and Dr. Towers had observed that it is very common for seriously ill patients to survive until an important event has transpired. Dr. Bahnson emphasized the importance of helping patients to set goals as an integral part of treatment—getting them to think about the next mountain to climb. He wondered whether oncologists influence the course of disease by their attitude of negative expectancy. Dr. Solomon suggested that it would be interesting to study the health outcome of patients of more sensitive/intuitive physicians. I added that I had never met an oncologist who couldn't point to at least one case which had defied his or her prognosis. What is significant is to pursue the reasons behind the improvement.

Dr. Ronald Glaser of Ohio State University, who served in an advisory capacity to the Task Force, viewed the step-by-step process of mapping nervous, endocrine, and immune system interactions as paralleling the charting of the way the body generates energy. He said psychoneuroimmunologic research is where metabolic research was fifty or sixty years ago, only psychoneuroimmunologic mechanisms are more complicated. Therefore, psychoneuroimmunologic investigators ought not to be so hard on themselves for not being able to come up with all the answers right away.

As the Task Force evolved, the discussion turned to issues of research methodology—or the scientific precision necessary to carry out meaningful psychoneuroimmunologic research.

Dr. Fawzy viewed clinical research in psychoneuroimmunology as having suffered from too many measurement and treatment methods. Dr. Bahnson commented on the failings of psychological measurement methods, cautioning that many tests have been based on special groups, such as college students, who might differ from research populations. The basic problem, however, was that what people say is not always what they feel. What they may put on paper does not necessarily reflect what is observed clinically. For instance, Dr. Bahnson cautioned the group not to be seduced by the apparent absence of depression in bulimic patients in that they sometimes express their depression in purging. Even in an anonymous testing situation, there are always some answers that are dictated by a desire to be socially acceptable. This phenomenon is particularly noticeable when an interviewer is present. Dr. Bahnson advocated using "old-fashioned" techniques—purposefully ambiguous psychological measurement methods (aimed at eliciting unconscious emotional states and psychological defense systems) that require some interpretive information.

Dr. Towers noted the difficulty experienced by many malignant melanoma patients in expressing their emotions and in producing visual imagery. However, once the patients overcame the difficulty, their physical condition improved. Dr. Bahnson pointed out that research participants should be kept as homogeneous as possible because individuals who are at different stages of the same illness or who have different illnesses exhibit different emotional and psychological patterns that should be studied separately from one another.

With regard to physiological measurements, Dr. Solomon alerted the group to a possible physiological "lag" period. Years ago, he had a rheumatoid arthritis patient whom he evaluated weekly. Since the exacerbation of arthritic symptoms is very highly associated with emotional states, Dr. Solomon scrutinized his patient's interview material from the week prior to clinical exacerbations and remissions. Not finding anything, he went back a week further and found a dramatic association between changes in symptom status and events that had occurred two weeks earlier. Dr. Solomon speculated that in addition to immediately observable changes, there may be physiological alterations that trigger a *chain* of events.

Dr. Marvin Stein, of Mt. Sinai School of Medicine in New York, who was a member of our extended advisory committee, addressed himself to critical issues related to research methodology. He said that some researchers, in measuring the effect of stress on the immune system, tend to overlook the possibility that the observed disturbance may be but a step in a series of immunological responses. For example, as laboratory studies have suggested, there may be an initial and temporary enhancement of immunity followed by an extreme decline. Medical researchers know they have to be alert to "rebound phenomena" and therefore check their findings over a sufficiently long period of time in order to obtain a clear view of settled effects.

Dr. Stein also emphasized that, in research on the effects of stress, more attention should be paid to describing the specifics regarding the particular nature of the stressor. Substances influencing immunity, such as neurotransmitters, hormones, and neuropeptides, can be altered by the specific nature of a stressor. Dr. Liebeskind concurred, explaining that his work demonstrated that the body responds with different types of analgesia production (i.e., opioid and nonopioid) and corresponding alterations in immune responses and tumor growth according to the specific characteristics of the stress-inducing situation.

Dr. Fahey suggested that patients in early stages of malignant melanoma, for whom surgery is recommended, would be a good group to study because they typically do not receive radiation and chemotherapy, treatments that may impair the immune function. He added that the immune system may play a role in containing metastasis in malignant melanoma cases. The group agreed on the importance of conducting research to test immune functions of patients over a period of time (in addition to the routinely done measurements of immune cell numbers). This research should also evaluate immune cell function as affected by immune-regulating substances, such as interleukin-2.

Dr. Fawzy said that recruiting malignant melanoma patients for research might best be done through the John Wayne Clinic at the UCLA Jonsson Comprehensive Cancer Center because of the large patient flow. He saw this as representing a unique opportunity to evaluate comprehensively a brief program offering psychological support, medical education, and stress management training. The aim should be to improve the psychological states of the patients and their quality of life.

Eventually, the Task Force members embarked on many of these projects or helped to support them at the medical center. By sharing information about research efforts and findings, we were educating ourselves even as we developed new information. This is how we hoped the program would develop—as a clearinghouse for research and education.

Supplementing these activities of the Task Force was a series of special programs at UCLA itself—lectures at the medical school, a postdoctoral training program, a research award program, an international conference, and courses in psychoneuroimmunology. The monthly lecture series, called the "Seminar on Behavioral Neuroimmunology," brought world recognized scientists and provided access to reports on current research in the field of psychoneuroimmunology. Some of the visiting authorities stayed for several days, allowing faculty, postdoctoral fellows, and graduate students to meet with them on an individual basis.

The Task Force instituted a postdoctoral training program in psychoneuroimmunology in which two individuals per year have been supervised by Task Force members and trained in behavioral neuroscience and immunological research methodology. In addition, the program sponsored a scholar in residence who spent his sabbatical acquiring knowledge in psychoneuroimmunology at UCLA in order to facilitate research at his own university. Some of the research being conducted by these individuals is described in Chapter 18.

The research award program in psychoneuroimmunology was developed to provide annual grants to UCLA scientists who wish to conduct research on the interactions of the psyche, nervous system, endocrine system, and immune system. Applications have been systematically reviewed by members of the Task Force. The recipients, along with the postdoctoral trainees, have met monthly with Dr. Fahey to discuss their research and new developments in the field, expanding and refining their interdisciplinary experimental capabilities. Another goal of these meetings has been to build a psychoneuroimmunology network. Dr. Clemente believed that such a communication network would elevate psychoneuroimmunology as an important area of science at UCLA.

The Task Force sponsored a major international conference to review the state of the art in psychoneuroimmunology and its role in an integrated medical philosophy. The conference articulated goals within the field in order to increase the quality and relevance of research. Individuals invited to the conference represented experts in neuroscience, immunology, psychology, psychiatry, medical philosophy, and science writing—people who have been noted for their integrative thinking and high-caliber work. A unique element of the conference was its focus on the philosophical and medical implications of psychoneuroimmunology; and, based on an integration of new research findings, its exploration of a theoretical framework for understanding and studying health and disease.

Individual members of the Task Force initiated courses in psychoneuroimmunology at UCLA. Dr. Fahey offered an advanced seminar for graduate students, faculty, postdoctoral fellows, and individuals doing psychoneuroimmunologic research. The course, entitled "Interactions Between the Immune and Nervous Systems," focused on methodological issues in neuroendocrinology, immunology, and psychology. Dr. Weiner coordinated a course for second-year medical students in which actual patient interviews were used to teach the art of thorough history-taking, and to explore the relationships of mental and emotional states, bodily functioning, and health. Dr. Fawzy organized an intensive psychoneuroimmunology course for third- and fourth-year medical students, providing them with exposure to both clinical and laboratory opportunities. Dr. Margaret Kemeny (postdoctoral fellow in psychoneuroimmunology) and Dr. Yehuda Shavit (then a postdoctoral student in Dr. Liebeskind's laboratory, now at The Hebrew University in Jerusalem) co-taught a seminar in behavioral neuroimmunology for both graduate and undergraduate students at UCLA. Dr. Towers coordinated full-day seminars on psychoneuroimmunology and the use of guided imagery in the treatment of patients. He also directed an ongoing series of monthly panel discussions on health issues, called the Medicine and Society Forum, for the medical school, the UCLA community, and the general public. I was invited to lecture at general meetings of medical students.

Our purpose was to get the students to ponder and probe the pathways between the directed use of intelligence and the organs of the body affected by illness or challenge in general—not *just* the intelligence, of course; all the aspects of consciousness and emotion that help to define the human condition. With the formation of the Task Force, we had taken a giant step forward.

16

ON THE DEFENSIVE

If I can look back on my "highs" during the years at the medical school, I have no difficulty in identifying the low point. It came in June 1985.

When I arrived at my office, Carol Prager, my secretary, said that all hell was breaking loose.

"*Time* magazine and *The New York Times* are trying to reach you. The *New England Journal of Medicine* has just published an article discrediting the notion that attitudes can make a difference in combating illness and they want a comment."

I found it difficult to get what she was saying into focus. Even before I could respond, the telephone rang.

"It's *Time* magazine again," Carol said. "Maybe you'd better take it."

The voice on the other end of the phone asked me if I had seen the article in the *New England Journal of Medicine* by Dr. Barrie R. Cassileth and her colleagues at the University of Pennsylvania medical school.

I said I hadn't and asked about the thrust of the piece.

The *Time* reporter on the phone said the *NEJM* article concerned a survey of 359 patients suffering from advanced cancer.

Patients were asked to identify their emotional states and attitudes toward the disease. The high death rate (75 percent) was interpreted by the authors as proof that those patients who had positive attitudes fared no better than those who did not. It was suggested, in fact, that the positive attitudes may have done harm because they led to feelings of personal failure when they didn't result in improvement or cure.

"That isn't all," the *Time* reporter said. "Let me quote the heart of the piece: 'Our study of patients with advanced, high-risk malignant diseases suggests that the inherent biology of the disease alone determines the prognosis, overriding the potentially mitigating influence of psychosocial factors.' What the article concludes is that you are wrong in believing that attitudes can affect medical treatment."

I could feel the ground slipping under my feet.

"Mr. Cousins, do you have a comment?"

"I would like to see the entire article first."

"Don't you have any reaction at all to the paragraph I read you? It's not out of context."

"The paragraph you read refers to high-risk, advanced cancer," I said. "In those cases, even the most heroic medical treatment may not override the biology of the disease. But high-risk cancer accounts for only a small fraction of cancer cases. For the rest, doctors want to throw all their forces into the battle—whether medical or psychological. Anyway, I really ought to see the full article."

"There's also an editorial in the magazine by one of the editors that is even more critical of the kind of ideas you've been associated with. It makes three points. First, that some members of the medical profession have been too ready to accept the belief that mental states are important factors in the cause and cure of disease. Second, that the medical literature contains very few scientifically sound studies of the relation, if there is one, between mental state and disease. Third, that the trouble with positive attitudes is that the patient is apt to feel a personal sense of failure if, despite those attitudes, the disease should continue on its downward course. Would you care to comment?"

"Same answer. If you wish, I'll be glad to phone you after I read the pieces in full."

"I'm not sure we have enough time. We're on deadline right now."

Matters were to get worse before they got better. The copy of *NEJM* containing the article and the editorial had not yet arrived at the UCLA Biomedical Library or any of the individual offices we called. It was a full day before the issue of NEJM was at hand.

Some newspaper reports seemed to indicate that I thought laughter was a substitute for competent medical care. Nowhere had I advocated laughter as an *alternative* to traditional medical attention. In fact, I had originally written the *Anatomy of an Illness* for the *New England Journal of Medicine* for the express purpose of correcting the notion, given currency in some newspaper accounts, that I had laughed my way out of a serious illness. I had referred to laughter as a metaphor for the full range of the positive emotions, including hope, faith, love, determination, purpose, festivity, and a strong will to live. I had also written about the value I attached to the partnership with my physician and the support he gave me for the experiment I proposed to see whether laughter actually helped to combat illness.

As I said earlier, I had reported that Dr. Hitzig was as fascinated as I was by the fact that the comic films I was watching in my hospital room were helpful in reducing my pains.

The experiment we had used, as mentioned earlier, involved a simple sedimentation-rate test. Since my illness was marked by an adverse sedimentation rate, it might be significant if laughter affected that rate. We therefore took sedimentation-time readings before and after laughter produced by comic films and discovered that the physiological effects of laughter were in fact real. But neither Dr. Hitzig nor I contended that laughter took the place of everything involved in traditional medical care—only that it could be fitted into it.

The article that appeared in *Time*, like other newspaper and magazine pieces, interpreted Dr. Cassileth's article as a scientific repudiation by the medical profession of the notion that attitudes can affect the course of disease in general. As one after another of these press clippings came to my attention, I sank lower and lower, feeling that six years of hopes and work at UCLA were going down the drain. Certainly, all those in the medical profession who were dubious about the original propositions could now feel confirmed in their view that what went on in the head was not involved in the treatment of illness.

As matters turned out, I need not have felt so desolate. I began to receive calls and letters from physicians and medical researchers strongly supporting the idea that a strong will to live and the positive emotions in general could make a difference in the patient's total situation. Most of these calls and letters, however, expressed the hope that we would be able to provide scientific evidence to support our position.

The most surprising—and welcome—call of all came from Dr. Cassileth herself.

"I've been terribly disturbed by some of the interpretations of our study," she said. "They have taken a point we were making about extreme cancer cases and have applied it to the treatment of *all* illnesses—and this is not what I believe or what I intended. And even in the extreme cancer cases, the way those patients think about their illness can affect the quality of their lives. I wish there were some way to counteract the misinterpretations of the article."

Out of that conversation came something highly useful. We agreed to issue a joint statement. It would demonstrate not only that there was no fundamental difference between us but that we strongly supported certain basic ideas. Dr. Cassileth and I prepared separate drafts for such a statement and then combined them. When released to the press, the statement carrying both our names took this form:

> The current public controversy over the relationship between emotions and health has placed the authors of this article on opposite sides. Much of the controversy, however, has its origin in serious misunderstandings of our basic positions. What concerns us especially is that these misunderstandings can produce public confusion and may cause harm to patients who are trying to mobilize their resources in the fight against disease.
>
> The confusion grew out of press reports concerning the article "Psychosocial Correlates of Survival in Advanced Malignant Disease?" written by Barrie R. Cassileth and colleagues and published in the *New England*

Journal of Medicine. Some of the reports and comments incorrectly interpreted the study's results to mean that positive attitudes have no value in a strategy for effective treatment of illness.

Cassileth's study, however, was not concerned with disease in general but with advanced cancer in particular. Cassileth wrote: "Our study of patients with advanced, high-risk malignant diseases suggests that the inherent biology of the disease alone determines the prognosis, overriding the potentially mitigating influence of psychosocial factors."

This means that in advanced cancer, biology overwhelms psychology. It does not mean that emotions and health are unrelated. It does not mean that emotions and attitudes play no role in the treatment or well-being of ill people.

In any case, high-risk cancer accounts for a very small percentage of all illnesses in the United States. The fact that positive attitudes or emotions cannot be expected to reverse or cure untreatable cancer does not mean they have no value in the large majority of illnesses. Indeed, positive attitudes of patients not only can enhance the environment of treatment, but can have a beneficial effect on the quality of life of patients. Physicians have always believed that a strong will to live helps a patient's chances in combating serious disease.

In an analogous fashion, Norman Cousins' work has been grossly simplified. His *Anatomy of an Illness*, first published in the *New England Journal of Medicine*, and his public statements concerning the complex relationship between mental attitude and physical health have been reduced in some quarters to the absurd notion that laughter can cure cancer.

Cousins used laughter as a metaphor for the full range of the positive emotions, including hope, love, faith, a strong will to live, determination, and purpose. He also stressed the importance of the patient-physician partnership in effective medical care.

We hope the following points will dispel the confusion, as well as clarify our points of view. Rather than being diametrically opposed, we share a common understanding and perspective.

- Emotions and health are closely related. It has been known for many years that negative emotions and experiences can have a deleterious effect on health and can complicate medical treatment. Not as well known is the connection between positive attitudes and the possible enhancement of the body's healing system. This relationship is now the subject of study at a number of medical research centers.
- It is likely that numerous emotional and physical factors, many of them yet to be delineated, influence health and disease, probably in different ways for different individuals. There is no single, simple factor that causes or cures cancer and other major illnesses.
- Even where positive attitudes and a good mental outlook cannot influence the physical outcome, they can and do affect the quality of life. Few things are more important in the care of seriously ill patients than their mental state and the general environment in which they have to be treated. Unfortunately, human beings are not able to exercise control over all of their biological and disease processes. Therefore, they should not be encouraged to believe that positive attitudes are a substitute for competent medical attention.
- Feelings of panic are not uncommon among patients in learning that they have cancer. Panic is itself destructive and can interfere with effective treatment. The wise physician, therefore, is mindful of the need to combat feelings of panic and emotional devastation.

The reciprocal mind/body relationship is complex. We must be aware equally of both the potential power and the limitations of attitudes in their effects on health and disease.

The Cassileth controversy continued to produce lively discussions at meetings of the Task Force. Members of the group felt that interpretations of the article bypassed decades of scientific evidence showing that psychological factors could be involved in the cause and progression of cancer. Numerous researchers had also shown that the *course* of cancer can be predicted by psychological factors.

For example, Drs. H. Steven Greer and Keith W. Pettingale of King's College School of Medicine and Dentistry in London, England, studied women who had undergone breast cancer surgery. They reported that survival without recurrence of cancer was more common among patients who responded with "fighting spirit" than among patients who expressed helplessness or hopelessness. Their findings suggest that positive attitudes had relevance not only as a method of determining prognosis but as an integral part of a strategy for combating the disease.

The connection between cancer and the expression of emotion was studied by Dr. Lydia Temoshok of the University of California at San Francisco, School of Medicine. Her structured interview with patients measured emotional, behavioral, physical, and mental reactions to events. These measurements revealed that malignant melanoma patients whose attitudes and emotions were active instead of passive exhibited better immune function and slower tumor growth.

Drs. Sandra M. Levy and Ronald B. Herberman, of the University of Pittsburgh School of Medicine and the Pittsburgh Cancer Institute, whose work is mentioned in Chapter 6, found that depressed cancer patients tended to have poorer NK-cell activity and greater likelihood of tumor spread. These findings were confirmed in a subsequent study of breast cancer patients. The investigators suggest that psychological factors could provide an index to the biological condition.

Both Dr. Temoshok's and Dr. Levy's studies linked emotional inhibitions to impairment in immune activity. This may explain the link between emotional suppression, such as passivity or stoicism, and the progression of cancer.

Dr. Leonard R. Derogatis and associates of The Johns Hopkins University School of Medicine, conducted psychological tests that might predict the length of survival of women with metastatic breast cancer. They reported that long-term survivors had higher scores for negative emotional expression than did short-term survivors. The patients' own oncologists described the long-term survivors as being significantly less "well adjusted" to their illnesses than short-term survivors.

Dr. G. Nicholas Rogentine, Jr., and colleagues of the National Cancer Institute recruited patients who had been successfully treated for malignant melanoma. The patients were asked to rate the amount of "adjustment" they required in order to cope with their illness. Participants who reported that they reconciled themselves to their illness were *more* prone to recurrence than those who resisted the idea of adapting to cancer.

The Task Force discussions about Dr. Cassileth's study emphasized the need for precise measurements in dealing with psychological factors. Dr. Solomon explained that only one of the measurements used in the Cassileth study had been found to be relevant to the progression of cancer. Dr. Liebeskind cautioned that in laboratory research radically different results can be produced depending on the procedure. Dr. Bahnson said that his own research showed that cancer patients do not always recognize or identify their feelings. Since cancer patients tend to report lower levels of emotional distress than their observed behavior or measurements of unconscious affect would indicate, Dr. Bahnson said it was essential in the study of cancer patients to measure unconscious psychological states using interviews or tests that allowed for more free association, such as story completion and picture interpretation.

The joint statement did much to correct the misimpressions that followed the original article. Even more reassuring to us were the results of a national survey of cancer specialists undertaken at UCLA. It tried to address the same issues as the Cassileth survey of patients. Ms. Linda Chilingar, statistical consultant, was in charge of the project.

We received responses from 649 oncologists who reported their opinions on the importance of various psychosocial factors, such as will to live, hopefulness, and coping ability. The physicians also commented on the value they attached to their interactions with patients. Those views were based on their treatment of more than 100,000 cancer patients.

We were tremendously heartened not just by the large percentage of replies we received but by the actual results. More than 90 percent of the physicians said they attached the highest value to attitudes of hope and optimism. There was overwhelming agreement on the proposition that a strong will to live, confidence in the physician, and emotional support from family and friends created an atmosphere conducive to effective treatment.

Comments accompanying the questionnaires reflected the view that the physician plays multiple roles in the treatment of cancer patients—diagnosis, prescription of specific treatment, emotional support for patient and family, improving the patient's quality of life and, in terminal cases, helping the patient to die in dignity and peace.

The commentaries by physicians also suggested the absence of any conflict between scientific and psychological factors in the care of cancer patients. Respect for the inherent biology of the disease does not preclude respect for the significance of patient attitudes and emotions. Far from disparaging the place of the positive emotions in the treatment of high-risk cancer, the oncologists as a group assigned considerable emphasis to the determination of patients to confront the challenge of serious illness. Such patient attitudes were considered second in importance only to early detection of disease and strict patient compliance.

The question concerning the willingness of the patients to comply with the prescribed treatment received a top "importance rating" of 93 percent. The comments accompanying the questionnaire tended to regard willingness to comply as a reflection of confidence in the physician. In this respect, such confidence was viewed as an essential aspect of effective medical care.

An equally significant fact emerging from the survey was that the oncologists as a group not only did not resist or resent a spirit of active and responsible involvement by their patients in a strategy of treatment but felt that such involvement helped to create an environment for optimal treatment. In this sense, the oncologists did not counterpose psychosocial factors against medical factors but favored a combination of the two.

Table 1 presents the general results of the survey.

TABLE 1
Relative Importance of Psychosocial Items in Contributing to Increased Longevity

RANK	BIOPSYCHOSOCIAL ITEMS	VERY OR MODERATELY IMPORTANT	SLIGHTLY OR NOT IMPORTANT
1	Seeking medical treatment at an early stage in the disease process	96%	4%
2	The patient's determination to comply with your recommendations	85%	15%
3	The patient's willingness to assume a proper share of responsibility for his or her own health and recovery	83%	17%
4	A positive approach to the challenge of the illness	80%	20%
5	Strong will to live	79%	21%
6	Confidence in you as the physician	77%	23%
7	Emotional support from friends and family members	75%	25%
8	The patient's ability to cope with stress	73%	27%
9	The patient's hopefulness concerning treatment outcome	68%	32%
10	The patient's willingness to communicate and share concerns with you	66%	34%
11	The patient's involvement in creative and meaningful activities	63%	37%
12	Strong religious or spiritual convictions	49%	51%
13	Previous successful experience with a life-threatening situation	41%	59%

Table 2 presents responses to questions in regard to patient-physician communication.

TABLE 2
Importance Attached to Physician-Patient Interactions

RANK	PHYSICIANS' SUPPORT SERVICES	ALWAYS OR OFTEN	RARELY OR NEVER
1	Answering the patient's questions about the disease, its treatment, side effects, and possible outcomes	99%	1%
2	Making sure that the patient clearly understands the explanation of the medical treatment procedures	99%	1%
3	Encouraging the patient to develop an attitude of hope and optimism concerning treatment outcome	95%	5%
4	Adjusting treatment plans to enhance compliance when the patient exhibits noncompliance	88%	12%
5	Directly counseling family members	87%	13%
6	Continuing to serve as primary physician when the patient receives supplementary treatment at another facility	85%	15%
7	Providing referral to social support groups	83%	17%
8	Providing the patient with educational materials about cancer	81%	19%
9	Helping the patient to develop methods to improve the quality of his or her life	74%	26%
10	Assisting the patient in determining which of his or her coping mechanisms are most productive, and helping to activate them	62%	38%
11	Providing referral to psychological counseling services	57%	43%

The survey asked the physicians how they felt about attending to the emotional needs of their patients. Most physicians said they wanted to be available to communicate with patients about all aspects of treatment.

One hundred ninety-three physicians offered comments on issues raised by the questionnaire. For example, most of them regarded emotional support as an important factor in the treatment of malignant disease. There was general agreement that physicians should help patients to improve their quality of life, regardless of the predicted length of survival.

Forty-two physicians offered the opinion that psychosocial factors have some effect on longevity. These physicians took the position that positive attitudes could have a beneficial effect on the patient's physical response to the disease. Five physicians qualified this premise, stating that positive attitudes were mainly useful for certain forms of cancer (e.g. breast, prostate, squamous cell carcinomas), and that the more aggressive cancers overwhelm otherwise positive psychosocial effects.

Some physicians pointed out that certain categories of patients respond differently to treatment. Some asserted that less intelligent patients worry less, are more compliant, and therefore tend to do well; others suggested that more intelligent patients make greater use of support services and are better able to commit themselves single-mindedly to a positive outcome. Nine physicians shared the observation that patients with a negative attitude (depressed, bitter, frustrated, despairing, recently retired, or recently failed in a life endeavor) generally experienced a rapid progression of their disease.

Fifteen physicians commented on the desirability of good patient-physician communication. It was suggested that the physician should encourage a realistic approach to the disease, being careful to discuss the illness as a challenge rather than a death sentence. Several respondents were concerned that unwarranted optimism would lead to feelings of guilt and inadequacy in the patient. Others commented on the importance of treating each patient as an individual, and remaining available to the patient at all times during and after treatment.

17

TRIUMPH WITHOUT VICTORY

Dr. James Maloney, then chief of surgery, UCLA School of Medicine, invited Ellen and me to his home one night to meet Mrs. Burton Bettingen, a benefactor of the university. Mrs. Bettingen had attended a lecture series on world affairs of which I was moderator and was eager to discuss some of the materials presented during the series. When I met Mrs. Bettingen, I was impressed not just with her knowledge of current events but with her lively sense of humor.

Dr. Maloney had also told Mrs. Bettingen of my work in combating despondency in patients.

Several days later, Mrs. Bettingen telephoned to say that her family physician, Dr. Clarence Hunter, was a patient at the UCLA hospital and she hoped I might try to cheer him up.

The request was not unusual. About one-fourth of the persons I am asked to see are physicians themselves. Emotional needs of seriously ill persons are not confined to people outside the medical profession. Physicians, especially, seem to reach out for an extra component beyond medical care. They recognize that organized knowledge can carry them only so far and that their own resources of body and mind have to be put to work as well.

I told Dr. Hunter that I had no magic formula for unlocking those inner resources but that I had brought with me some reports of new research from the files of our Task Force in Psychoneuroimmunology. These papers reported on emotional factors as activators of the immune system. At the very least, the ability to control depression connected to serious illness sometimes tends to bolster the body's cancer-fighting capacity. There was also evidence that a decrease in depression could create an internal environment conducive to medical treatment. We went on to discuss other matters; we swapped stories and reminisced about our contrasting experiences. After a while, Dr. Hunter seemed comfortable and was in a good mood.

I went to see him several times in the hospital and was pleased to learn from the nurses and residents that he was responding superbly to treatment. The progress continued day by day. One of the residents said he believed Dr. Hunter would make it all the way.

Mrs. Bettingen had a wide circle of friends; and hardly a month passed when I didn't receive a request to meet with one or another. As I got to know Mrs. Bettingen I realized I was in the presence of one of the most interesting people I had ever met. Her home was located atop a high ridge overlooking most of Beverly Hills and Century City. Inside, the house had the appearance of the world's largest antique shop. Mrs. Bettingen had highly developed tastes in objets d'art; she enjoyed living with favorite items in her collection.

Her father was Burton Green, one of the key figures in the early history of Beverly Hills. She related fascinating accounts of her father's vision in laying out the streets of the new city. He had conceived of Beverly Hills as a country place for the busy and wealthy people of Pasadena, some twenty miles to the east. Pasadena was becoming increasingly beset by smog in contrast to the clear bracing air in Beverly Hills with its long vistas and high ridges.

Mrs. Bettingen took out some albums and proudly showed early photographs. Rows of baby palm trees, laid out in long regular rows on the sides of the streets, were just coming up.

(Now they rise fifty feet or more, a magnificent colonnade of verdant sentries.)

"My father had extraordinary vision," she said. "We would ride horses up and down the narrow canyons and he would describe what he felt would one day become one of the most beautiful living communities in the United States. He spoke about building a great public school system with the kind of education that would enable young people to place a proper value on life and on their own potentialities. One of the first things Dad did was to build the Beverly Hills Hotel, which has become world famous.

"Growing up was a lot of fun. Seeing Dad's vision become reality was constantly exciting. Dad named me Burton after himself. They expected a boy and were prepared to call him Junior. When I arrived, they kept the name. I was one of three girls."

A few days later, I noticed that Mrs. Bettingen had circled an item in the newspaper about homeless young women. When I asked her about the news item, she was nothing if not animated in her reply.

"It's absolutely shocking," she said. "We're supposed to be one of the wealthiest and most caring societies in the world. Yet people are sleeping in hallways and on the streets. Some of them are young girls. It's something that should fill us with shame. What can we do about it?"

"I'm not sure," I said. "Maybe the problem ought to be dramatized in a way that captures public attention and makes it a major issue."

"And how do we do that?"

"Maybe a program or two on public television would be a beginning."

"Good idea," she said. "Now, how do we do that?"

"Let me talk to Bill Kobin about it. He's the new president of KCET. I got to know him some years ago when he was program director of National Educational Television and I was chairman of the board."

"Great idea," she exclaimed. "I'll be glad to pay for it."

The next day I telephoned Bill Kobin, who accepted the proposal at once. When I asked how much it would cost, he said somewhere between $150,000 and $200,000.

This was far higher than I thought was within the anticipations of Mrs. Bettingen. I almost hesitated the next afternoon to report to her about the conversation with Bill Kobin. When she asked about it, I shifted uneasily in my seat.

"Come to the point," Mrs. Bettingen said sternly. "What figure did he give you?"

"Something in excess of one hundred fifty thousand dollars," I said in a low voice.

"That's not so bad," she said. "Tell him to go ahead. I'll send him the check. Norman, do you hear me?"

"Yes, I hear you."

I notified Mr. Kobin and the arrangement was completed.

It was a good program. I don't know what effect it had. But one thing was certain: Mrs. Bettingen was enjoying herself.

"There must be other things we can do," she said brightly.

Thus began six weeks of exploration and adventure in unorthodox giving—most of it designed to help individuals and all of it anonymous.

Then came a Saturday afternoon I will never forget. She said she was discovering possibilities that gave her a sense of purpose and were bringing her rewards. She turned to me abruptly and asked:

"Do you have any dream—any wild hope of something you would like to do?"

"I'm full of wild dreams," I said.

"But is there any one dream that is really doable?"

I didn't have to stumble for an answer.

"Certainly," I said. "I'd like to see our Program in Psychoneuroimmunology at UCLA become the finest in the world. We've already come a long way towards this goal through a gift from Mrs. Joan Kroc."

"What would an expanded program do?" she asked.

"It would act as a world clearinghouse for information about the way the mind can have physical effects on the body. It would stage conferences at which leading experts would come together to report on their research. We would enlarge our present efforts of research at UCLA and elsewhere."

I then reviewed the research at UCLA and elsewhere made possible by Mrs. Kroc.

"Do you have any idea what a permanent program would cost?"

"At a guess, I think we could make a pretty good start with five million dollars."

Mrs. Bettingen didn't hesitate.

"Patty," she called out to her secretary, "Come here and bring my checkbook."

Patty did so.

Mrs. Bettingen made out a check to the medical school for $5 million and handed it to me.

"Tell the school to hold on to the check for a couple of days," she said. "I'm doing some switching in my bank accounts."

On Monday morning, I went directly to Dean Mellinkoff's office. I said nothing but handed him the check. His expression indicated he thought I was indulging in one of my spoofs.

"This is all very interesting," he said. "Is the check real?"

"It's real enough," I replied.

He sat down slowly and I told him the story of Mrs. Bettingen.

"Nothing like this has ever happened in the history of the university," he said. "Are you sure the check is good?"

"Well, she did ask us to hold it up for a couple of days."

The next day I conveyed the thanks of the dean and the chancellor to Mrs. Bettingen.

"I'm not through with you yet," she said with mock severity. "Jim Maloney has been talking about needing a new surgical facility at UCLA. It will cost about $11 million, and I want him to have it. I think he'll be pleased, don't you?"

I grinned and nodded. I felt as though the heavens had opened up and a magic wand had appeared in the sky.

"You know," she said, "this has been a lot of fun. We have something in common. We've both discovered the power of the pen. Incidentally, the other day you said something about a new facility on the west side of the UCLA campus that might be available for your Program in Psychoneuroimmunology. Maybe we can drive out to see it—say on Wednesday morning?"

On Wednesday at about 11:00 A.M. I telephoned Patty Brown to ask if Mrs. Bettingen was ready to be picked up. The line was busy. I waited fifteen minutes and phoned again. Still busy. It was now almost 11:30. I began to be concerned when the busy signal continued. I telephoned Ellen at the house and asked her to run over to Mrs. Bettingen's—it was only five or six minutes away—and see if Mrs. Bettingen wanted to keep the appointment.

A little while later Ellen phoned to report that the door to Mrs. Bettingen's bedroom was still closed. Patty Brown felt we should not awaken her; Mrs. Bettingen might not have slept well during the night. Perhaps that was why the phone was off the hook.

Some time later, Patty telephoned. Her voice seemed to come from far away.

"Mrs. Bettingen's gone," she said.

"Where did she go?"

Silence.

"Patty, are you there?"

"Yes. I mean Mrs. Bettingen's gone. I mean, she passed away."

I rushed over to the house. Dr. Clarence Hunter and Dr. Arthur Samuels had already arrived.

"I think she became ill very suddenly," Patty said.

Dr. Samuels completed his examination.

"She had a heart attack," he said when he emerged from the bedroom. "Apparently death was very sudden."

Dr. Hunter proceeded to examine Mrs. Bettingen in Dr. Samuels' presence. He concurred with Dr. Samuels' findings.

Death is a natural habitat for irony. Mrs. Bettingen was discovering powers she never fully realized she had. Just as she was embarking on a new life, she was struck down by a heart attack. Camus might have termed it an existential event with echoes of Greek drama. She learned she could make a difference in the world. It gave her a sense of purpose and fulfillment.

At least, this is the way I like to think about Burtie Bettingen.

News of Mrs. Bettingen's gift spread quickly. The inevitable question kept coming my way: How were we going to spend the money?

We were only a few weeks away from the end of the three-year period made possible by Joan Kroc. In fact, we had just sent Mrs. Kroc and Mrs. Benes a progress report covering the period of the grant. Mrs. Bettingen's gift put us in a good position to develop an endowment. But interest rates had dropped sharply. A few years earlier, the prevailing interest rates were in the vicinity of 12 percent. Now, with current interest rates only 7 percent or so, we would have only enough interest to cover about half of our annual budget.

After Mrs. Kroc learned about the Bettingen gift, she suggested that, rather than spend the money for operational purposes, we might wish to consider using it as an endowment. She offered to bolster the endowment with an additional gift of $3 million. She had read our comprehensive report and was pleased by the progress we had made toward our original goal—mounting and supporting research to show that positive emotions could make a difference for the better in confronting the challenge of illness—or any challenge, for that matter.

Joan B. Kroc had taken a big gamble when she supported our work at the outset. She didn't shrink from taking risks. She recognized that many foundations had a tendency to play it safe—too safe.

Alexis de Tocqueville, in his classic study, *Democracy in America,* a century and a half ago, was fascinated with the sense of responsibility for the well-being of society exhibited by some individuals in the United States. In no other country did he find so strong a tendency to return wealth to the people. It was a form of ultimate power—but power implies knowledge, courage, initiative, moral imagination. Joan Kroc, like Burtie Bettingen, exemplified these qualities. I was determined as never before to justify their confidence in us.

18

BELIEF BECOMES BIOLOGY

As the work of the Task Force progressed and the results arrived of research projects we had designed or helped to fund, we began to perceive the grand outlines of connections between body and mind that were far more intensive and extensive than had generally been supposed. Our excitement mounted with the report of each research project. It was apparent that the obsession that brought me to UCLA was not as fanciful as some had suggested or as I had sometimes feared. Pieces of evidence were falling into place and were creating a coherent and consistent design. This chapter contains a summary listing of many of the studies and preliminary findings. A kind of mosaic was emerging in which we are still finding the pieces, many still missing.

We were especially interested in a number of studies showing the way the human mind converts ideas and expectations into biochemical realities. Perhaps the most dramatic example of these studies concerned the phenomenon of the placebo. Medical researchers have been increasingly fascinated by the fact that many people, when told what to expect in the way of a medication's effects, will actually experience those effects even though the "medication" is replaced by a pill containing a totally innocuous substance, perhaps sugar or saline. Why should the anticipation of a physical effect bring about actual physical change? And if anticipation or attitudes have a role in creating physical change, how can that knowledge be used to enhance medical treatment or promote good health?

A number of placebo studies have shown significant bodily changes as the result of mental processes. For example, Dr. Ronald Katz, chairman of the Department of Anesthesiology at the UCLA School of Medicine, reported a series of observations involving patients who were informed that headaches were a complication of spinal anesthesia. At the last minute the patients were told that the choice of anesthesia had been changed from spinal to general. Despite this fact, all the patients experienced the symptoms that went with spinal anesthesia. Internists and neurologists who examined the patients reported "classical examples of spinal anesthesia headache."

Dr. Katz's study was similar to the placebo experiment of Dr. J. W. L. Fielding, Department of Surgery at Queen Elizabeth Hospital in Birmingham, England. In accordance with the informed consent procedure, 411 patients were told they could expect hair loss from the chemotherapy about to be administered. Thirty percent of the patients unknowingly received placebos instead of chemotherapy and suffered hair loss even though the pills they had taken contained no medication.

One of the world's leading experts on behavioral medicine, Dr. Neal Miller of The Rockefeller University and Yale University, has emphasized that what is most significant about the placebo response is the proof it offers that thoughts or expectations can be converted into physiological reality. Dr. Miller's work makes a good case for the strategic and careful use of placebos in the treatment of patients.

That expectations can have a specific effect on the immune system has been demonstrated by Drs. Robert Ader and Nicholas Cohen of the University of Rochester, who have shown that the immune system can be trained, or "conditioned," to respond to a neutral stimulus (placebo), as first reported in 1926 by S. Metal'nikov and V. Chorine of the Pasteur Institute in Paris. Drs. Ader and Cohen found that the administration of an immune-suppressing drug and placebo together "conditioned" the immune system to respond to the placebo alone after the drug was discontinued. They also found that, by alternating the administration of real medication and placebos, thus "conditioning" the body's physiological response to the placebo, the conditioning effects of a drug can be increased. Also, side effects and dependence (as well as expense) may be reduced. Drs. Ader and Cohen have demonstrated drug conditioning to be so effective that dosages of medication that in themselves are not strong enough to reverse disease can nevertheless slow its progress.

Dr. Ader and Dr. Anthony Suchman have extended their research to the treatment of human disorders. In a recent study funded by the Task Force, the researchers found that when hypertensive patients were taken off drugs and treated with placebos, they maintained normal blood pressures longer than patients who were taken off active medication and treated with nothing. The implications of this research on important pharmacological treatments are most encouraging.

Anesthesiologist Dr. Henry K. Beecher of Harvard Medical School, who pioneered in this field, observed that the greater the pain or anxiety, the more effective the placebo. He noted that the usefulness of *any* given drug is a combination of its chemical ingredients and the confidence of the patient that it will work. This accounts for the effectiveness of placebos in ameliorating a variety of disorders ranging from angina pectoris and gastrointestinal disorders to fever and the common cold.

Dr. Jon D. Levine of the Department of Neurology at the University of California at San Francisco Medical School, studied postoperative dental patients. He found that pain relief from expectations linked to placebos actually triggered the body's own anesthesia, or endorphins. Such research suggests that the psyche, sometimes regarded as a separate entity from the rest of the body, is involved in the regulation of bodily processes.

The natural conclusion emerging from placebo research, of course, is that belief affects biology. The response of individuals to the world around them, touching off hopes or fears or joys or despairs or expectations in general, has physical reality. This knowledge is integral to treatment of illness. It doesn't mean that medical treatment should be supplanted by psychological or emotional approaches, but that the effective reach of the physician can be expanded by awareness of emotional and psychological factors involved in the cause of disease and in a comprehensive strategy of treatment.

Adjacent to the evidence offered by placebo research were studies supported by the Task Force showing the power of suggestion. Dr. Zvi Bentwich, of the Kaplan Medical Center in Rehovot, Israel, studied the recovery patterns of three groups of patients recovering from surgery for the repair of a hernia. One group listened to tape-recorded messages of a rapid recovery, another was told they would receive absent healing treatment by psychic healers, and one had only standard medical treatment. Measures of recovery being evaluated included the need for pain medication, ability to get up from bed, healing of the surgical wound, and ability to return to regular activities and work. Results of the study at this writing indicate that the belief of patients can be a factor in recovery.

Dr. Steven E. Locke of Harvard Medical School, who specializes in behavorial medicine and diseases related to disorders of the immune system, has studied the power of suggestion, using techniques such as hypnotism in relation to skin test immune responses. Thus far, he has observed a *generalized* link between specific hypnotic suggestions and immune reactions. Dr. Locke co-edited *Mind and Immunity: Behavioral Immunology*, a bibliography of scientific papers in the field of psychoneuroimmunology, published by the Institute for the Advancement of Health.

The surroundings of medical testing and care, including the attitudes and manner of the caregivers, carry more weight in the outcome for the patient than is generally recognized by busy and hard-pressed medical staffs. Dr. Fawzy, of our Task Force, with whom I was associated in the project to increase the coping abilities of cancer patients, recognized this need and was involved in research projects related to the environment of health care. The first of these projects called for hospital personnel to be trained in the special requirements of caring for seriously ill patients. If emotional strain could intensify the disease, freedom from such strain should be regarded as an essential aspect of treatment. It might be supposed that sensitivity to the emotional needs of patients and families is a standard feature in the training of all hospital personnel. In actual fact, equal emphasis is not given to emotional and physical requirements of patients in many hospital training programs.

Dr. Fawzy's second project involved a training program in communication skills for medical students. The program also sought to make not just students but interns and residents aware of their own feelings and reactions to patients.

The matter of the environment as it relates specifically to testing was the subject of research by Dr. Anthony Reading, UCLA psychologist. Dr. Reading was interested in the chapter in my book *The Healing Heart* dealing with contrasting results on my treadmill test according to changed circumstances. The first test, conducted in the usual manner, resulted in highly adverse readings. Even before going on the treadmill, I had been apprehensive about the procedure. A friend of mine had died on a treadmill. I had no sense of control. My muscles were being moved involuntarily. Did it make no difference whether I was exercising or being exercised? What, actually, was being measured? It was no surprise to me that I had "flunked" my treadmill test. The measurements indicated very little cardiac ability to sustain exercise. I went directly from the treadmill room to the UCLA track. I had myself attached to a "Holter" device—a small machine that records heart action on tape that is later played back into a cardiograph. The main difference between the Holter and the treadmill cardiograph is that the Holter measures heart action under normal circumstances. It records the ability of the heart to sustain exercise that is connected to your own volition. I subjected myself to at least ten times as much exercise with the Holter as I had taken on the treadmill, then turned in the Holter tape for the cardiograph reading. It showed none of the adverse indications that were recorded in the treadmill room.

Dr. Reading wanted to use this experience as the basis for an extensive experiment in which cardiac patients would be tested under varying circumstances. He embarked on a series of studies dealing with the effects of physiological states on cardiac testing. He contrasted the results obtained under ordinary treadmill testing with careful psychological preparation to enable the patient to take part in the procedure with feelings of confidence and control. By changing the atmosphere of the test, he was able to show improved systolic blood pressure response during early stages of the procedure.

The initial results of another UCLA-supported project by Dr. Alan Rozanski of the UCLA School of Medicine, who is also director of Cardiac Rehabilitation at Cedars Sinai Medical Center in Los Angeles, seem to confirm the hypothesis that the standard exercise electrocardiogram test produces a more discouraging cardiovascular profile than the monitoring of heart function during daily life circumstances. Dr. Rozanski's work has also provided firm evidence that mental stress causes substantial and usually imperceptible myocardial ischemia (insufficient blood flow to the heart), which is a major manifestation of arterial blockage. Both Dr. Rozanski's and Dr. Reading's work confirmed our contention that the circumstances of medical testing could affect the outcome.

Perhaps no subject I have investigated in my ten years at the UCLA School of Medicine comes more frequently to my attention in one context or another, or continues to grow in significance in terms of health care, than that of the patient-physician relationship. The Task Force supported two studies at UCLA involving slightly different approaches to the nature of communication between patient and physician. Based on the premise that doctors exert an influence on health outcomes of their patients by shaping feelings and expectations, Drs. Sheldon Greenfield and Sherrie Kaplan of the UCLA School of Public Health, conducted four separate studies on the health status of patients with ulcer disease, hypertension, diabetes, and breast cancer. In each study, half of the patients were given general information regarding self-observation and care activities immediately prior to their regularly scheduled appointment, while the other half of the patients were given a twenty-minute training session on how to take a more active role in their medical care. The training sessions included an explanation of each patient's medical record; the basis upon which his or her treatment ws determined; and coaching on question asking, negotiating, maintaining "focus" on medical care issues, and on minimizing self-consciousness and inhibition. Drs. Greenfield and Kaplan found that increased patient control, more expression of affect by doctor and patient, and greater information provided by the doctor in response to patient questions, were related to better patient health status as measured by audiotapes of office visits, questionnaires, and physiological measurements.

Research such as Drs. Kaplan and Greenfield's project suggests that a more active patient role helps to foster a greater sense of control over illness, better health outcome, more rapid recovery, and greater compliance with treatment. Dr. Rose Maly of the UCLA School of Public Health utilized a simple technique to improve patient interactions with their physicians. The study observed corresponding attitudinal, behavioral, and health status changes in patients, as well as attitudinal changes in physicians. One group of patients was given their medical charts to review and completed a list of questions that the doctor discussed with the patient. The results from this group were compared with a control group that received only a health fact sheet to review and a patient suggestion form to complete.

Preliminary results indicate a significant improvement in the functional status of those who experienced the enhanced interaction with their physician—the benefits having their greatest impact on individuals over age sixty. As Drs. Kaplan and Greenfield demonstrated, a very simple effort to improve patient-physician interactions can result in significant improvement in a patient's condition.

The part played by the human mind and spirit in the healing process was explored in a variety of ways. One of the projects we helped to support was carried out by Dr. Herbert Benson of Harvard Medical School, who developed considerable data on the value of what he called the "relaxation response." To some people, the term has connotations of a person flopping down on a beach, arms akimbo, with Hawaiian guitars strumming in the background. Such, however, was not what Herbert Benson had in mind. When he spoke about relaxation, he was not thinking of transient or surface tensions that are readily banished by something as simple as a few minutes of repose at the seashore. He was proposing systematic ways of dispatching deep disturbances or fears that were having adverse effects on human health.

Dr. Benson has been assessing the psychological and affective benefits of seven weeks of relaxation response training (consisting of meditation relaxation combined with guided imagery). Patients undergoing this training have reported significant improvements in quality of life, including increased vigor and fighting spirit, along with a decrease in hopelessness, tension, depression, anxiety, and somatization. Dr. Benson also found relaxation response training effective in moderating the adverse physical effects of chemotherapy.

Preliminary results have shown that the gains made as a result of participation in the relaxation response group endure over time and that patients with more advanced cancer who rated highly on "fighting spirit" survived significantly longer. The significance of Dr. Benson's work was the proof it offered that intelligence and free will can help to combat existing health problems and prevent future ones. The role, therefore, of the mind in creating a useful design for living was both clear and emphatic.

The effects of relaxation and mental stress on the immune system were the focus of a study by Dr. Arthur A. Stone of the State University of New York at Stony Brook. In order to study these effects, Dr. Stone had a group of students listen to a relaxation tape for four sessions while a control group read magazines. Each of the groups was then given a stressful mental task to accomplish. Although the relaxation training was effective in inducing relaxation during the session, it did not have any aftereffects on psychological or immunological measures. However, the mental stress tasks caused measurable increases in cardiovascular and psychological stress, and lymphocyte stimulability was significantly lower for one hour immediately following the stressful tasks. The study confirmed findings of other research that mental stress can affect the immune system, in this case over a brief period of time. It also underlined the importance of our major quest, which was to find the evidence that the opposite of stress could also have an effect on the immune system.

The effectiveness of relaxation response training was also discussed in Chapter 3 in regard to the work of Drs. Glaser and Kiecolt-Glaser.

Two approaches to the use of visual or guided imagery have been supported by our UCLA Program in Psychoneuroimmunology. One by Dr. Bernard Towers, of our Task Force, involved a lecture series for medical students on the use of guided imagery to enhance the medical treatment of patients by helping patients to mobilize their internal healing resources. This series also dealt with medical ethics, providing students exposure to medical issues requiring sensitivity and awareness on the part of the physician. The other study was initiated by Dr. Inge B. Corless, chair of the Department of Care at the University of North Carolina. Dr. Corless developed a film to foster relaxation and positive mental states by means of visual imagery. The film is being used to facilitate patient recovery from illness.

We have been able to help fund projects in the Los Angeles area that provide unusual approaches to the emotional support of cancer patients, or to the attitudes of children toward illness. Devra Breslow, director for Special Programs at the UCLA Jonsson Comprehensive Cancer Center, introduced the Art-as-Therapy program for hospitalized cancer patients. This program provides art therapy, guided by artist-psychotherapists, to facilitate the expression of feelings and the resolution of illness-related problems. Some years ago, an organization in Los Angeles called We Can Do! was begun by a cancer patient who recognized the need for group support systems in meeting the physical and psychological needs of cancer patients. Patients affiliated with We Can Do! typically live well beyond the survival times predicted by their physicians. With funding from our UCLA program, the group has produced a new manual on stress management and coping skills for their participants. As mentioned in Chapter 10, a nonprofit organization called Humor/x, founded by Mrs. Bea Ammidown Miller, supplies "Laugh Wagons" to health care centers. The portable units carry humorous books, audiotapes, and videotapes to brighten hospital settings and encourage patients to take a more active role in their recovery.

Under the direction of Dean Lewis Solmon, the UCLA Graduate School of Education has instigated the development of a prevention-oriented health program for the University Elementary School. In contrast to the "Just Say No" themes of recent health education campaigns, this program, "Say Yes to Wellness," emphasizes physical, mental and social health, as well as the development of decision making and personal effectiveness skills.

As mentioned elsewhere in this book, medical literature rarely refers to the human healing system. The question of the role played by various systems in bodily repair and healing has been addressed by a number of the research studies supported by our program at UCLA. Theodore Melnechuk, visiting lecturer at the School of Medicine of the University of California at San

Diego, reviewed all the scientific evidence regarding the influence of the psyche, nervous, and endocrine systems on bodily repair systems. His report summarizes what is known about the healing process. One of the conclusions of the report is that the positive emotions are associated with improved wound healing. It is possible that these emotions alter production of hormones, neurotransmitters, and opioids that interfere with the various steps in the healing process. Distress has also been shown to damage DNA (genetic) repair. Resulting cell mutations are then genetically reproduced. These conclusions reflect the observations of a variety of clinical studies.

Drs. George Solomon and John Morley, of the UCLA School of Medicine have been involved in projects related to the interaction between psychosocial factors and the immune function in healthy elderly persons. Contrary to some expectations, they found that the immune systems of elderly people who are in good health are as functional as, and in some ways actually superior to, those of healthy persons half their age. They also found that natural killer (NK) cell stimulability was associated with emotional "hardiness" in response to life events, and was diminished by worry about anticipated stressful events. Their research suggests a stronger relationship between psychological factors and immune function in healthy elderly than in healthy young persons—a result supported by the work of Drs. Steven Schleifer, Steven Keller, and Marvin Stein mentioned in Chapter 6. They found that depression suppresses immunity to an increasing degree with advancing age. In a series of follow-up studies on factors that could help bolster the immune systems of elderly persons, Drs. Solomon and Morley demonstrated that the NK-cell activity of old and young subjects was comparably enhanced following a bout of vigorous physical exercise and that the body's opioid system (which includes endorphins) seems to be involved in causing these changes. Having demonstrated an increase in NK-cell activity as a result of physical activity, the investigators have been studying the effects of both stressful and nonstressful mental exercise on the immune system. They are evaluating whether mental exercise, such as challenge, either positively or negatively affects NK-cell activity in elderly subjects.

In another study, Dr. Solomon and associates evaluated psychological characteristics common to six AIDS patients who lived far longer than expected and who have been able to function at a remarkably high level. Dr. Solomon's findings suggest that positive attitudes and adaptive coping can bolster components of the immune system in a way that may help to offset the devastation resulting from the loss of helper-T cells. Dr. Solomon's research would seem to underline the importance of the advice we give to cancer patients: Don't deny the diagnosis; just defy the verdict that is supposed to go with it. This project was the basis for an intensive study of a larger group of AIDS patients. Again, the study found positive attitudes, emotional fortitude, taking an active role in one's own health maintenance, and attention to one's own needs to be related to relatively good immune measures including what might represent compensatory increases in some categories of immune cells. On the other hand, emotional distress was accompanied by negative immune function measures in persons with AIDS.

Research has shown that whereas some individuals infected with the human immunodeficiency virus (HIV) have rapidly deteriorating immune systems, others do not. In an effort to determine whether psychological factors play a role in the rate of imunological deterioration, Dr. Margaret Kemeny, postdoctoral fellow in the UCLA Program in Psychoneuroimmunology, has been documenting the positive and negative psychological characteristics of healthy HIV-infected homosexual men and observing immune and health changes over time. Preliminary results suggest that HIV-infected men who are depressed but not recently bereaved show immune changes that favor the development of AIDS. These individuals, who have fewer and more poorly functioning helper T cells, will be followed long-term to see whether or not they are more prone to developing AIDS. This research provides an extension to Dr. Solomon's findings that positive attitudes and emotional fortitude are characteristic of long-survivors of AIDS.

The question of the impact of the emotions on the immune system is being studied in relation to genital herpes patients. Dr. Leonard Zegans of the University of California at San Francisco School of Medicine, is studying ways in which combinations of brief and chronic emotional states influence the psychological, immunological, and recurrence status of genital herpes patients as compared to herpes-free participants. Dr. Zegans's previous research with Dr. Margaret Kemeny of UCLA has suggested that emotional adjustment is associated with certain immune processes and herpes outbreaks. These studies, combined with Dr. Kemeny's research on HIV-infected individuals, will add to the understanding of how positive and negative emotions influence immunity and health.

Another member of our Task Force, Dr. Claus Bahnson, has done pioneer work in the development of emotional and psychological profiles in connection with cancer patients. His recent study, supported by our program, compares the emotional, attitudinal, and quality-of-life status of cancer patients experiencing emotional conflict resolution through psychotherapy as compared to those receiving educational literature and attending informal question-and-answer sessions. He has also been measuring whether psychological and emotional improvements are reflected in neuroendocrine and immunologic changes, and in the course of disease. Although the data has yet to be analyzed, Dr. Bahnson has observed that the psychotherapy group participants have greatly benefited from their experience.

A number of other researchers, with the help of our program, are looking at aspects of the relationship between the emotions and the immune system. They include a study by Dr. Shlomo Breznitz, of the University of Haifa in Israel, to determine what biochemical changes accompany hope or discouragement in soldiers given varying information about the duration of a stressful task; research, by Dr. Eli Sercarz of the Department of Microbiology and Immunology at UCLA, on how emotional reactions based on social and environmental factors affect T-cell activity; research on the effects of psychotherapy on rheumatoid arthritis patients, by Josephine Rhodes of the University of Louisville; research on the impact of relaxation and imagery training aimed at eliciting positive emotions on the immune status of breast cancer patients, by Dr. Jane Taylor of the Stehlin Foundation for Cancer Research in Houston; and two studies relating to the biochemical pathways whereby stress impacts the development of ulcers and the immune system—by Dr. Robert L. Stephens of the Department of Psychiatry at UCLA and by Dr. Michael Irwin of the Department of Psychiatry at the University of California at San Diego School of Medicine, respectively.

The amount of attention given by researchers to the connections between the emotions and the immune system is indicative of the importance now associated with this relationship. Another area that has received less attention, but is of particular significance to patients and their families, is that of the impact of serious illness on the lives of those involved. One of the studies supported by our UCLA program dealt with the life-style adjustment problems of patients suffering from chronic illness. Dr. Alfred Katz of the UCLA School of Public Health found that within six months after diagnosis, 20 percent of the patients in his study experienced divorce or separation largely due to spousal disbelief in the illness (in this case chronic systemic lupus erythematosus) and poor adaptation to changing needs and roles subsequent to illness. Regular employment suffered heavily, as did routine chores, social activities, and psychological well-being. Data from self-help discussion group participants indicate that regular participation in such groups serves to boost psychological morale, self-esteem, and coping abilities.

In a study focusing on the life-adjustment needs of adult heart-transplant patients, Dr. Deane Wolcott of the UCLA Department of Psychiatry and Biobehaviorial Sciences has been collecting information from heart-transplant candidates on their quality of life before receiving a transplant and for two years after transplantation. Measurements of the quality of life include health status, routine daily activity, psychological well-being, social relationships, cognitive function, job function, and leisure activity participation. The documentation will also assist Dr. Wolcott in determining what psychosocial characteristics might be used most effectively to predict the overall success of treatment.

One of our Task Force members, Dr. John Fahey, created a psychoneuroimmunology laboratory, which has been providing immunological analyses for research and training for postdoctoral fellows, visiting scholars, and other research associates at UCLA. The techniques developed by Dr. Fahey are being used to evaluate immunological changes associated with disease, mental states, or alterations in the nervous, endocrine, and immune systems. His laboratory has been helpful in a number of studies connecting belief to biology.

The Task Force took pride in undertaking and supporting

all these research projects, many of which showed specific connections between mental states and bodily processes, and which were subsequently reported in medical journals. In addition, at the time of this writing, a number of other research undertakings were initiated and are summarized in Chapter 23 under "Ongoing Studies in Biochemical Pathways."

19

FUNCTIONAL AGE

Five basic misconceptions dominate and indeed disfigure much of the thinking about human health:

1. That almost all illnesses are caused by disease germs or other external factors.
2. That illness proceeds in a straight line unless interrupted by outside intervention in one form or another.
3. That pain is always a manifestation of disease and that the elimination of pain is therefore a manifestation of a return of "good" health.
4. That what goes into the mind has little or no effect on the body (and vice versa).
5. That old age is connected to numbers—that it begins at sixty-five, that mental and physical abilities beyond that point fall off significantly, and that society is justified in mandating retirement on the basis of age.

Other chapters in this book address the first four misconceptions. This chapter seeks to deal with the fifth.

Few things about life in America are more striking than the emphasis on youth. "Staying young" is not only a national obsession but the basis of a large part of the national economy. Obviously, there is nothing wrong in the desire to be active and attractive. What is terribly mistaken is the notion that these qualities are associated almost exclusively with youth. The society as a whole pays a fearful price for this error, depriving itself of a valuable resource—the skills and experience of a large segment of the population.

Society has yet to catch up with the fact that the prolongation of life has been accompanied by the prolongation of productive capacities. Medical research has provided scientific verification that age by itself does not automatically diminish brain power. As mentioned in Chapter 18, George Solomon and John Morley, in a project funded by our Task Force on Psychoneuroimmunology, have been able to show that the immune functions of healthy elderly persons compare favorably with those of far younger people. Conventional medical wisdom has held that one of the inevitable characteristics of advancing age is a progressive weakening of the immune system. The research of Solomon and Morley shows that this is not necessarily so.

The following excerpts are drawn directly from the report of their study:

> *The immune system is not uniformly affected by the aging process. For example, total numbers of white blood cells, lymphocytes, and granulocytes, as well as phagocytic function of neutrophils and the complement system do not change appreciably with age. . . .*
>
> *Healthy elderly subjects demonstrate enhanced NK (natural killer-cell) activity both at baseline and after stimulation with beta-endorphin, as compared to a younger population. Analysis of the Leu-11a and Leu-19 lymphocyte subgroups indicates that this enhanced activity may reflect an increase in the proportion of cells with NK activity in the healthy elderly. We suggest that this difference between healthy young and old subjects may represent a cohort of effect of increased immune surveillance in those surviving to old age . . . .*
>
> *The elderly also demonstrate an enhanced stimulation of NK activity by beta-endorphin. It is sug-*

gested that beta-endorphin, which is released from the pituitary in concert with ACTH in response to stress, provides a potential link between stress, altered immunity, and diseases related to dysregulation of the immune system. Support for this hypothesis comes from our finding that beta-endorphin plays a role in the stimulation of NK activity associated with exercise. "Hardiness," a constellation of personality traits and coping patterns reflecting commitment, control, and challenge that has previously been found to be related to health, correlates with stimulation of NK cell activity. Since IL-2 may be important for augmentation of NK activity in vitro, *augmented activity would mediate stronger surveillance against malignant and infectious disorders. The strong correlation found between emotional "hardiness" and IL-2 stimulation of NK cells both in elderly and in young subjects suggests at least one mechanism by which hardiness influences maintenance of physical health.*

Future studies will be needed to determine whether both superior psychological and immunological functions are necessary to maintain health in old age.

Medical research is discovering that high determination and purpose can actually enhance the working of the immune system. It is not surprising, therefore, that Dr. Solomon and his colleagues should have developed the evidence that there is no automatic correlation between advancing age and a deterioration of the immune system. The ability of the human body to turn back illness is one of the wonders of the world. Indeed, the more we know about the connection between mind and body, the greater the prospect that we can consciously put it to work for our greater good.

These studies indicate, therefore, that elderly persons who have managed to maintain reasonably good health do not experience the deterioration of the immune system usually associated with their age. The main problem represented by advancing age may well be connected to negative expectations.

And, of course, health is not the only significant immune galvanizer in the aging; there's security and sufficient income too. Since human beings tend to move along the path of these expectations, morose thoughts about the aging process tend to set the stage for incapacity and disability. The big challenge confronting both society and the elderly today is the acceptance of new realities. We are reminded of the "barrier" of the four-minute mile. Once Roger Bannister accomplished the feat, in 1954, almost every major track competition has seen runners come in well under four minutes. As soon as a new reality was demonstrated, new capacities came into being.

It is being increasingly recognized that the sixty-five-year-old dividing line is arbitrary and absurd. A vast range of new possibilities now comes into view. Many companies are revising their policies on the retirement age. Some organizations go even further, making exceptions in individual situations. UCLA provides a series of extensions, one year at a time, for faculty members who exhibit undiminished capacity. Having been the beneficiary of such exemptions several times over, I find it difficult to argue against the new policy.

As mentioned in Chapter 1, I was diagnosed as having tuberculosis at the age of ten. I suspect that my physicians at that time would have hesitated to predict that I would be fully functional at the age of seventy-four. For that matter, the same might be said of physicians who turned me down for insurance at the age of thirty-nine or specialists who gave me a diagnosis of ankylosing spondylitis (inflammation of the spinal joints) at age forty-nine. I have no trace today of the threatened paralysis; I am able to engage in vigorous sports without physical limitations. I mention these facts only to bear personal witness to the ability of the human body to respond to challenge.

One of the most farsighted books of the twentieth century is Elie Metchnikoff's *The Nature of Man: Studies in Optimistic Philosophy*, published in 1906. It is interesting to note that it remained for a Russian to write a book about the organization and function of the human body that corresponds to the dominant American political and philosophical strain, beginning with Jefferson, Franklin, and Adams and extending through to Emerson, Fuller, William James, Pierce, and John Dewey. That strain is connected to the idea of human perfectibility. Not perfection, for no one can define fully the nature of perfection or its outer limits, perfectability as represented by creative growth, betterment, and the pursuit of human potentiality. Metchnikoff accepted no fixed ceiling over human biology. At the time he wrote his book, average life expectancy in the United States was in the mid-forties. He could look back over the course of human development and reflect on the thousands of years it took for average life expectancy to double from twenty. Today we look back and contemplate the fact that it took only seventy years to add the same number of years to the life span that it did in the preceding thousand years or more. The human organism has responded magnificently to improved nutrition and modern sanitation.

One recalls that H. G. Wells, when asked to list the greatest developments leading to extended longevity, put sanitary plumbing in general and flush toilets in particular close to the top. The remark was not as capricious as it might seem. As a historian, Wells was aware of the epidemics caused by faulty sanitation. The sanitary disposal of human wastes, in addition to safe drinking water, may not be recorded in the history books among the great achievements of civilization, but there can be no doubt of their contribution to human longevity—exceeded perhaps only by the advent of antibiotics, which have played a major part in enabling medicine to become a science as well as an art.

In any case, Metchnikoff's expectations about the ability of human beings to break through supposed age barriers have turned out to be correct. The increase in life expectancy from forty-seven to seventy-four in only three-quarters of a century is one of the great developments in history. Within less than two generations, most of the major killing diseases that have plagued humankind have been conquered. Poliomyelitis, smallpox, tuberculosis, leprosy, diphtheria, and gastroenteritis—scourges throughout history—have either been eradicated or brought within controllable limits. Today, cancer and cardiovascular diseases are the main threat to longevity, with AIDS potentially an even greater danger; but it would be unhistorical to believe that they are altogether beyond the reach of the kind of scientific intelligence that has achieved spectacular successes in the past.

Dr. Nathan Shock, of the National Institute of Aging, has

calculated that the conquest of cardiovascular disease and cancer would increase average age by at least six years—a development that alone could give many people now alive octogenarian status.

What are the chances that cancer, diabetes, and the diseases related to clogged arteries can be conquered in the foreseeable future?

First, knowledge is fast accumulating on the connection between life-style and serious disease. The way the emotions are processed by the brain, affecting almost every function of the body, including digestion, respiration, circulation, heart, hormones, and immune system, can serve as an important guidepost in any program of disease prevention. Similarly, adequate exercise and rigorous attention to nutritional requirements of the human body have a place in a comprehensive strategy for maintaining health and in combating illness.

In terms of specific advances, Hodgkin's disease and some forms of leukemia are being brought under increasing control. Exciting new possibilities in the treatment of prostatic and other tumors appear to be opening up through the use of antioxidants, selenium, and retinoids. New ways of controlling blood sugar, such as through infusion pumps for insulin, give promise of coping more effectively with diabetes. Perhaps the greatest advances of all may be represented by new approaches to arteriosclerosis. Low-density lipoproteins, one of the main markers of clogged arteries, are being attacked with increasing effectiveness not just through exotic drugs but through dietary change and exercise.

Finally, recognition is growing that rapid aging is a disease that can be brought under greater control. Students of the aging process are scrutinizing the connection between cellular change and general bodily deterioration. New knowledge of the role of DNA in health may lead to important genetic advances with a corresponding deferral of the effects of aging.

The recent advances in medical science are accompanied by a corresponding upgrading in the daily life-style, with greater respect for the nutritional needs of the human body and increased control over the circumstances of life. The resultant prospects for meaningful longer life are being substantially enhanced. Adding another twenty-five or even more years to present life expectations is a realistic possibility during the next half-century.

Life expectancy is now moving toward eighty years of age, with the prospect that it may go as high as ninety in the early decades of the twenty-first century. At a time when the national economy is dependent on increased and improved productivity, the supposedly "elderly" represent a prodigious resource ready to be converted into a national asset.

Medical science has devised a wide battery of tests to determine functional capacity. There is every prospect that these tests can be further developed and refined. Such tests can be used as a basis for a fundamental change in the society's policy toward the elderly. In any event, it is necessary to get away from sovereignty of numbers—whether seventy or eighty or ninety—and apply individual considerations as a basis for rational and responsible decisions in determining "functional age."

20

THE SOCIETY OF CHALLENGERS

Early in 1987, Jolly West telephoned to tell me something I had been waiting years to hear. He said that Dr. Fawzy Fawzy was in his office with an exciting proposal that sought to provide scientific evidence of the connection between the positive emotions and improvement in the condition of seriously ill patients. Would I come over?

Within ten minutes I was listening to Fawzy, who was highly regarded at UCLA for his psychiatric work with cancer patients. He was in his early forties, carried himself with great dignity, and was soft-spoken and kindly in manner.

"We've been talking about ways of *proving* that the mind-set of cancer patients may actually affect their ability to psychologically and physically deal with their disease," Fawzy said. "I've been discussing this general idea with Don Morton and John Fahey. We've drawn up a protocol for a research project. It's pretty far-reaching; in fact, it's probably the most extensive research proposal of its kind. We've got a good chance of mounting the kind of research that can authoritatively show that a strong will to live and hopefulness can make a difference in the prospects of cancer patients."

Fawzy was saying exactly what I had been longing to hear.

Jolly West winked at me and smiled.

"I thought you might be interested," he said.

"The idea is to work with a significant number of melanoma patients," Fawzy continued. "I don't have to tell you that malignant melanoma is a particularly insidious form of cancer. We would divide these patients into two groups of approximately equal size—one group to serve as the control, the other as the experimental.

"Both these groups would receive whatever medical attention is required—medical or surgical. The only difference between the two groups is that the research patients would receive psychological, educational, and intellectual help and support. We hope you will want to do for the research group the kind of work you are doing for cancer patients at the V.A. hospital. We want you to help them believe in themselves and their doctors. You can help them to cope, lift their spirits, give them a design for living. I think you get the idea. I'll be with you and will deal with whatever psychiatric problems come up. We'll work as a team. Don Morton and John Fahey have helped with the protocol. That will mean a great deal inside the medical profession."

Don Morton was the well-known oncological surgeon; John Fahey was one of the key national figures in immunology, and, of course, a member of our Task Force.

Fawzy showed me the protocol. It called for blood measurements of the melanoma patients for a sustained period—a year or more. Certain kinds of immune cells would be regularly measured during the course of the experiment. The patients would also be tested for levels of depression and anxiety on a POMS (Profile of Mood States) scale—a standard test for measuring a variety of mood disturbances—and on the PAIS (Psychosocial Adjustment to Illness Inventory), which measures psychological distress in response to physical illness. The POMS test is a listing of sixty-five adjectives describing a variety of moods. Individuals are asked to rate, on a five-point scale, the frequency with which they experience the mood. In the PAIS test, multiple-choice answers are provided for forty-six questions regarding health care attitudes, vocation, relationships, social activities, and psychological distress.

Psychological and immunological measurements would be

done at baseline, six weeks later, after six months, and at one year to see whether immune enhancement, if any, was purely temporary or whether something lasting was being built into their lives.

I asked Fawzy how the patients for both the control and research group would be selected.

"From among the people who come to our cancer clinic," he said. "We'll just describe, as matter-of-factly as possible, what we have in mind."

The protocol Fawzy showed me had an initial budget of $80,000. I couldn't think of a better way to spend some of the money given us by Mrs. Kroc.

It would be six or seven weeks before the project was fully in place. Meanwhile, I decided to experiment on myself. The scientific evidence was clear that deep and sustained depression actually reduces the number of cancer-fighting cells in the immune system. Such being the case, I naturally wondered whether the opposites of depression—joyous anticipations, etc.—could actually increase the population of immune cells.

I had two blood samples drawn from my arm. The first blood sample I used as a baseline; we took the second sample five minutes later. During the interval, I tried to put myself in a mood of pleasant expectations and, in general, to experience emotional well-being.

I realized that five minutes was a very short interval—so short, in fact, that it might be unreasonable to expect any significant changes. Yet I was concerned that, if we waited much longer, it might be said that intervening factors had affected the result. The advantage of a five-minute interval, of course, was that, if in fact a change did take place, the cause-and-effect relationship would be established.

After taking the first blood sample, I tried to imagine the very best thing that could happen to our world—and then tried to make it seem as real as possible. I imagined what a wonderful planet we might have if the Soviet Union and the United States had rational foreign policies. I tried to visualize all the benefits that might accrue to the human race if the madness at the top subsided and the leaders of the major nations broke away from the accumulation of weapons that, if used, could change the basic conditions of life itself. I tried to imagine the changes for the better that might flow out of the realization that true security depended on the creation of workable world order.

I confess I found this train of thought both energizing and stimulating; after about five minutes, we took the second blood sample. The results couldn't have been more gratifying. In just five minutes there was an average increase of 53 percent in the various components of my immune system—from a low of 30 percent in the precursor NK (natural killer) cells to a high of about 200 percent in the antibody-coated T cells.

This was just a single case, of course, and it had no standing in medical science. But it was also true that what happened in my case might conceivably happen to others. One instance does not qualify as science, but it helps to nourish a theory and to set the stage for systematic research. I looked forward to the work with the melanoma patients, fortified in the belief that we were not fumbling in the dark altogether.

A good name for the project on which we were to embark was the "Society of Challengers." Meetings of the Society were to take place on Saturday mornings and run through noon. The patients would arrive an hour or so earlier to have their blood drawn for their immune measurements. We would divide the research patients into units of no more than fifteen each. The units would meet sequentially; that is, we would try to complete the research with one unit before starting with the next. If it became necessary later to schedule more than one meeting on the same day, we would have shorter sessions but hold to Saturday mornings.

With Dr. Fawzy's permission I brought a talented psychotherapist, Evelyn Silvers, with me to our first meeting—and also to subsequent ones. Her previous experiences with patients, I felt, would be an important resource for the entire group.

Our Saturday morning meetings were held in one of the conference rooms of the UCLA Neuropsychiatric Institute. What was most encouraging was the responsiveness of the patients and their obvious eagerness to meet our expectations.

I suggested to Dr. Fawzy, based on my meetings with cancer patients at the Sepulveda Veterans' Hospital, that we have the melanoma patients sit in a large circle. In this way they would create a physical unity; everyone would see everyone else. At the first session, Dr. Fawzy explained the background of the project and reviewed the scientific evidence on which our hopes were based. He stated that the meetings would be informal yet have some structure. Over the six weeks they would receive educational information about their cancer and about nutrition and would be taught various relaxation techniques as well as positive coping strategies and problem-solving techniques. Each week we would encourage them to talk about the good things that had happened since the previous meeting. We would not make this a hard-and-fast rule, of course—the patients could talk about anything they wished—but the hope was that we could create a source of mutual nourishment.

When Dr. Fawzy turned the meeting over to me, I suggested that we introduce ourselves as fully as possible, talking about our experiences, our fears, hopes, our perceptions of what led to this particular juncture in our lives.

The patients fell in with the suggestion and developed a momentum as they went along. Each person seemed to draw additional encouragement from what had been said. There was progressively less hesitation to talk about the relationships with physicians or about the emotional events that preceded the illness.

Donald, the first man who spoke (all names of patients are fictitious) had come in a wheelchair, his right foot massively bandaged and propped up on the leg support. He was the youngest of seven children. His father had been a professional baseball player for the St. Louis Cardinals. Later, his father became chief of police in St. Louis. He grew up with an interest in show business. He studied the drums under private tutors, then made the jump to big-name bands in the Catskill Mountains summer resorts. By the age of nineteen he was playing in leading hotels in Las Vegas and Los Angeles. In his mid-thirties he gave up drums for business. One day, he discovered a growth on his toe. The diagnosis of malignant melanoma sent him spinning into a depression, which in turn led him to this group.

Mary, well poised, well spoken, had had a varied career as newspaper reporter and editor, consultant to the food industry, and marketing manufacturer for a major food distributor. She grew up in a military family that moved about a great deal. Family life was not especially affectionate or emotionally demonstrative. Her interest in and knowledge of the food business was acquired from her mother, who worked as a food broker.

Before being diagnosed, Mary was under a great deal of emotional stress. In the first six years of her marriage she gave birth to a daughter, then had two miscarriages. Shortly after the second episode, her gynecologist discovered malignant melanoma. One of her main problems was that she had no one to talk to. She read a great deal in the medical press about melanoma; the more she read, the lonelier she became. She found it necessary to write a letter to her husband, to be opened after her death. It was a list of instructions about how to take care of their daughter.

Surgery removed the cancer and the prognosis was guardedly good. But her inability to reach out and have people reach back put her into deep depression. She said she had high hopes for her membership in our group and was especially grateful for the chance to meet with and talk to people whose physical and emotional problems were not too different from her own.

Joanne had recently experienced marital separation. The main thing wrong with the marriage was that she felt her husband had no real need for her. The relationship grew thinner and thinner; the decision to separate was inevitable. About that time she met and became attached to a man who had severe health problems. She emerged from one situation in which she felt completely useless only to be plunged into another in which she felt overmatched and overwhelmed. During this period of melancholy and mid-passage, she was attacked by an unknown assailant. She began to develop multiple symptoms. A complete medical workup at UCLA revealed that malignant melanoma had attacked her lungs and liver. Her first reaction, after all that she had been through, was that the melanoma was just another in a series of episodes that had to be faced. She came to our group, she said, because she felt instinctively that if she could just find an emotional haven, her natural optimism would pull her through. Even if her odds of recovery were only one in five, she decided she would be the one.

Andrew could look back on a rewarding upbringing. His father was a nationally known foundation executive. His mother met all the classical attributes—loving, nurturing, inspiring, supportive. In his second year at college, Andrew got married, continued at school part-time, gained employment in the defense industry working on antiaircraft. The country was then deep in the Vietnam War. Andrew, who opposed the war, was unhappy with himself for working in an antiaircraft plant. His marriage broke up, he remarried, and his new wife helped see him through college to a degree. After graduating from college, he obtained a job as a high school English teacher. Some students had complained to him about being emotionally harassed by another teacher. The teacher responded to his criticisms by accusing Andrew of having had intimacies with her students. Charges flew back and forth and made newspaper headlines. He was forced to resign from his job. Meanwhile, his wife left. He also lost his job at the school. A few months later, he noticed some swellings. The diagnosis of malignant melanoma was unequivocal. Andrew contemplated suicide, then decided that the most important lesson he had learned as a youth—summon all your courage in meeting big challenges—was an asset to be put to work. He underwent surgery three times. "If it doesn't work this time, you're history," one of the surgeons said about the third operation.

The surgery did work. His wife returned with their daughter, and both of them provided physical and spiritual support. He was rehired as a part-time teacher. It was at this point that he became part of the UCLA group, which rounded out all the positive things that were happening in his life.

Susan grew up in a poor neighborhood in Brooklyn, New York. In her early years, she stuttered badly but improved as she developed increasing confidence in her abilities as scholar, dancer, athlete. She married outside her faith to an accountant and real estate operator. Her great disillusion came when she discovered that her husband had taken all her savings and invested them under his own name in a real estate venture that ultimately failed. There were other abuses and breaches of good faith. The marriage broke up and Susan was left without a home. Just at that time, her mother received a cancer diagnosis, dying a few months later. Shortly thereafter, Susan herself became ill. The diagnosis was malignant melanoma. It was against this background that she learned about the UCLA group.

•

Brendan grew up in a large Catholic family in Missouri during the Depression years. He left high school to enlist in the navy, after which he got a job in earth and space technology. At the same time, he involved himself in volunteer social service activities. He had no difficulty in gaining the respect and affection of the people with whom he worked, either in social welfare or at the plant. Brendan experienced no particular emotional stress or turmoil prior to his melanoma. Surgery went off without difficulty. Brendan's plans for marriage were not set aside by the fact of his illness. His wife became a prime source of spiritual support.

When Brendan heard about the UCLA group, he applied for membership but was turned down because certain features of his treatment program interfered with the kinds of measurements that were being taken. But Brendan's social welfare background caused him to persist; he thought he might be helpful with the other patients. The doctors finally acquiesced to his request.

As the sessions of the Society of Challengers progressed, we discovered that the interest they took in one another mounted week by week. Relationships were formed that were continued outside our meetings.

The patients made known to Fawzy and me their expectation that the two of us would join the group in its autobiographical journeys. They seemed to be particularly curious about my experiences at the *Saturday Review* and would ask questions about our exchanges with noted writers. They seemed delighted when I told them about the abbreviated "fistfight" between Max Eastman and Ernest Hemingway in the office of Maxwell Perkins, their editor at Scribner's. Following the encounter, Eastman had come to the offices of the *Saturday Review* and poured out his side of the story. Two days later, Hemingway came to lunch with *SR*'s editors at the old Seymour Hotel opposite the offices of the magazine. He had his own version of the fight.

As might be expected, the accounts varied widely. Maxwell Perkins told us he had never intended that the two writers meet in the reception room. Hemingway had come very late, Eastman very early. At first, their conversation was polite, then veered in a competitive direction, with Hemingway boasting he had more hair on his chest than Eastman, and with Eastman taking him up on the challenge. The exchange rapidly escalated and became a chest-bumping and shoving affair. Despite the reports in the press, no punches were actually thrown, but there was enough contact to make for colorful first-page stories.

With the encouragement of the Challengers group, I also spoke of my own illnesses and my convictions about the role of the mind in contributing to an effective strategy of healing. I gave as much emphasis as I could to the major theme of my talks with patients—"Don't deny the diagnosis; just defy the verdict that is supposed to go with it." And I would talk about specific cases, with photographs, of patients who overcame the grim predictions of physicians and either conquered their illness or were able to prolong their lives by many years.

I didn't miss any opportunities, nor did Fawzy, for producing robust laughter. And, at Fawzy's request, I would put the entire group through the routine I had found useful in helping cancer patients to get over feelings marked by loss of control and helplessness. The routine was the one referred to earlier in this book—showing them how to move their blood around and then getting them to realize, as a result, that they were not barred from some measure of control over bodily processes that they had supposed were locked into the autonomic nervous system. We didn't have enough hand thermometers to go around, but when the patients pressed the palms of their hands against their cheeks and felt the great warmth, their sense of surprise and fascination was no less great than it was with the individual patients. They relished the evidence that they had the power to increase the surface temperature of their skin by ten degrees or more.

The main strategy at the meetings, then, was to give the cancer patients a sense that they were not lacking in resources of their own to throw into the fight. They realized that they could combine those resources with those of their physicians in meeting a serious challenge. They learned that hope need not be spurious or synthetic, and that a blazing determination and strong will to live were provable assets.

Week by week, the patients became less tentative, less uncertain, less weighed down by depression. As they gained in confidence, individually and collectively, they were able to play increasingly important roles in the lives of one another. During these weeks, the immune systems of the participants were being regularly measured. We were also measuring levels of depression.

As pointed out earlier, any serious illness has a tendency to produce melancholy and despair. A person who has suffered a heart attack, for example, has to come to terms with the meaning of the episode once the emergency phase of the attack is over. In cancer, this realization may be even more stark. The resultant depression is almost universal. But depression can produce profound physiological change, adversely affecting homeostasis, the biochemical balances of the body. Measurement of depression, therefore, was as significant in our study as the measurement of cancer-fighting cells in the immune system. When we compared the POMS measurements of the control group against the research group at six weeks, we found that the mean change scores for the control group showed a slight decrease, whereas the experimental group showed a significant drop in depression. At first, we thought the variation might be too temporary to have any significance. However, it not only continued but accelerated. By the end of six months, a substantial decline in the depression level of the research group was documented compared to an *increase* in depression for the control group. At six months there was also a dramatic difference in the PAIS scores, with the control group showing a slight decrease in psychological distress compared to a more marked decrease for the experimental group.

The reeducation of the patients, apparently, was having its effects. The growth in confidence; the increasing knowledge by patients about the nature of their own resources; the enhancement of life-style; the decline in feelings of helplessness—all these were reflected in the POMS and PAIS measurements of the research group.

Most exciting of all, however, was that the decline in depression was accompanied by an increase in certain immune cells, or activating forces, within the immune system. The conclusion was inescapable: If you can reduce the depression that almost invariably affects cancer patients, you can increase the body's own capacity for combating malignancies. This becomes especially important in view of the fact that chemotherapy, which

How Depression and Quality of Life Affect the Immune System

PROFILE OF MOOD STATES (POMS)
Mean Change Scores

TENSION-ANXIETY	POST-TREATMENT		SIX MONTHS POST-TREATMENT	
Control	0.15	P<.012*	0.04	P<.007*
Experimental	−4.06		−4.34	
DEPRESSION-DEJECTION				
Control	−0.58	P<.049*	0.04	P<.003*
Experimental	−3.89		−4.71	

PSYCHOSOCIAL ADJUSTMENT TO ILLNESS (PAIS)
Mean Change Scores

PSYCHOLOGICAL DISTRESS			
Control	(PAIS test not administered at this point)	−1.46	P<.043*
Experimental		−5.77	
TOTAL PAIS			
Control		−1.35	P<.007*
Experimental		−6.91	

QUALITY OF LIFE
Mean Change Scores

Control	−1.65	P<.024*	−0.23	P<.035*
Experimental	7.60		8.34	

IMMUNE CELLS (in the NK cell family)
Mean Change Scores

LEU 7				
Control	−0.85	P<.032*	0.04	P<.044*
Experimental	1.06		2.09	

LEU 11				
Control	0.50		0.12	
		P<.740		P<.014*
Experimental	0.80		2.89	

*P is the level of statistical significance between the experimental and control groups. P<0.05 is considered to be statistically significant. All statistically significant figures have been asterisked.

is often used in the treatment of cancer, can have deleterious effects on the immune system.

I have known some physicians to escalate the administration of chemotherapy—*aggressive chemotherapy* is the term—even as patients were suffering a drastic intensification of their symptoms. In most of these cases, the sharp increase in use of the powerful chemicals had catastrophic effects.

When I raised the question with oncologists about the wisdom and efficacy of this approach, they called attention to the physician's dilemma.

"A patient takes a dramatic downturn and it is clear that existing procedures are not enough," Don Morton told me. "The family becomes desperate. They come to the doctor and say, 'Please do something.' By that time most physicians have exhausted the other options—radiation and surgery, especially. And so they feel they have to do something—generally more of what they're already doing—and hope for the best."

It is reasonable, of course, to ask whether such treatment doesn't have the effect of punishing the patient for his or her illness. Physicians are struggling for better answers than they have now. And every once in a while the miracle happens.

In any event, while the issue is still open, the wise physician takes into account the need to invoke the fullest possible use of the body's own immune capacity. And that capacity can be affected by the extent to which the patient becomes free of feelings of helplessness and despair. It is apparent, therefore, that chemotherapy, to be used optimally, should be combined with a systematic program to combat depression. The evidence is fast accumulating that if the body is to be treated effectively, the wise physician will pay attention to what goes on in the mind.

Even more heartening, perhaps, than the proof of a connection between control of depression and an increase in immune capacity was the physical condition of the Challengers. By the end of the year, we didn't have to dwell on statistics for evidence of what was happening. We could clearly observe the improvement in the health of the members of the research group. There was hardly a person who didn't experience an upturn as well as an improved prognosis from the oncologists.

The most dramatic case of all may have been represented by Don, who had come to the first meeting with his foot massively bandaged and propped up in his wheelchair after a series of unsuccessful graft attempts following his malignant melanoma surgery. Within only four months, Don was out of the wheelchair and on crutches. After another three months, he was off crutches and was able to get around with a cane. At the end of the year, he was doing a jig for the entire group. His triumph didn't belong to him alone. Every member of the group could share in the elation.

Two young women whose weddings had been called off months earlier because of their melanomas were sufficiently improved to proceed with their marriages. We had a special party to celebrate their new status in life.

And so it went. There was little doubt in my mind that the research project that began with the protocol drawn up by Dr. Fawzy in consultation with Dr. Morton and Dr. Fahey had produced the evidence that I was looking for when I came to UCLA. This evidence, to be sure, had to be carefully evaluated, checked, and cross-checked before it could be presented to the medical community. Moreover, the research statistics in the Society of Challengers project had not yet been fully computerized. But the indications seemed good, very good.

The deep feeling of gratification I had over the result of the malignant melanoma study was enlarged as the result of another UCLA research project, described in the opening pages and elsewhere in this book. The aim here was to see the effects of varying emotions—especially feelings of joy—on the immune system. Dr. David Shapiro, professor in the department of Psychiatry, UCLA, and Ann Futterman, Ph.D. candidate, designed a most imaginative study involving professional actors. The underlying idea was that good acting requires that the actor actually "feel" the role he or she is called upon to play. If substantial evidence could be developed that certain emotions could be "ordered" into being, with corresponding physiological benefits, then physicians would have a basis of incorporating psychological strategies into their treatment.

The authors of the study applied to the Task Force in Psychoneuroimmunology for a grant. Their proposal was squarely in line with the central purpose of our Task Force. The main aim of the project was to test human ability to steer the emotions in the direction of psychological change. A checklist of positive and negative emotions was drawn up. If sadness was the designated emotion the actors would be asked to portray a scene involving great sadness. Then, by contrast, the actors would be given a scene calling for expression of deep joy or festivity.

The design for the study called for blood samples to be taken before and after each scene. The samples would be run through the flow cytometer for measurements of the immune cells. A wide battery of other tests was brought into play—galvanic skin response, blood pressure, effects on white and red blood cells, etc.

The Shapiro-Futterman research went forward along the lines described in the first chapter. At this writing, definitive results are not at hand because of the small number of actors involved so far. But even on that limited basis, the information yielded is highly significant and supports the results of other research reported in this book about the physiological effects of the emotions. What gives this particular study special value, of course, is its indication that, if patients can bring certain emotions into play, they may enhance the environment of medical treatment even as they may produce beneficial physiological changes.

21

OBSESSION REVISITED

Friday, December 2, 1987, was a banner day in my ten years at the UCLA medical school. It began with an excited telephone call from Dr. Fawzy.

"I've just seen the preliminary computerized statistics," he said. "Everything we have observed clinically has been confirmed by thousands of figures on various aspects of our research project. We're completely vindicated in our basic theory."

The elation in his voice was unmistakable. I told him to come over at once. When he arrived he quickly produced a sheaf of papers.

"The computer printout," he said, "shows the difference between the melanoma patients in the control group, who received only the prescribed medical and surgical treatment, and those in the research group, who received the prescribed treatment plus education in ways of coping psychologically with serious disease."

Dr. Fawzy spread his first batch of statistics on the table.

"We've now got the hard scientific evidence that emotions and states of mind make a real difference," he said. "Here are the comparative measurements that show marked differences in the immune systems between the two groups of patients. These figures are not just one-time evaluations. They continue over the period of more than a year. They show the unmistakably enhanced effects on the immune system of patients who were able to reduce anxiety about their illness and cope with life stresses more effectively. Here, look at this."

He pointed to a row of figures.

"These are the Leu-seven cells. They are one of the natural killer cell series in the immune system that help to destroy cancer cells. The mean change scores showed that the control group's cells had actually decreased while the experimental group showed the desired increase in these cells at six weeks. By six months the control group had managed to return to close to baseline while the experimental group had continued to increase their Leu-seven cells. This trend continues in many of the other important cell categories."

I thought of the marked improvement in the clinical condition of individual patients in the year since the project started—Don, who did a jig for the group at the end of one year; Joanne, who had not been expected to survive for more than a few months but whose improvement was so striking that she was able to proceed with her deferred wedding plans at the end of the year. These were only two of many. What was most significant about the computer printout Fawzy showed me was that the numbers represented real people who had a new claim on life.

"We've got to expect considerable resistance to these findings," Fawzy said. "Despite all the evidence, some physicians will persist in thinking that what goes on in the mind has nothing to do with illness or recovery. Our hope has to be that most physicians will carefully scrutinize the results of our study objectively and will recognize that it is possible to enhance the treatment effect by enlisting the hope, determination, and will to live of patients. Physicians would also come to accept that no conflict between physiological and psychological factors exists. They would then recognize the value of a combined strategy of treatment as integral to effective health care."

What seemed to me to be especially significant in the results of the study was the clear evidence of a relationship between depression and health. Patients who were liberated from depression and despair showed an increase in the number of cancer-fighting immune cells. This connection threw valuable light on the need for patients to participate actively with their physicians in a total program for combating serious illness.

In any case, the obsession that brought me to UCLA—a drive to find scientific evidence that the positive emotions could produce positive biological changes—had produced key elements of the evidence I was looking for. It was not only the UCLA Challengers project that demonstrated the way mind and body communicated with one another. Equally important were a dozen or more research efforts carried out elsewhere, some of which we had helped to fund. These undertakings helped to create a far better picture than we had had previously of the way attitudes and moods have their effects on illness and health.

This didn't mean that additional research was not necessary. Just the opposite: Significant research had only just begun. But at least and at last it was possible to talk scientifically about the importance of hope and the will to live as essential parts of a comprehensive strategy of medical care.

22

REPORT TO THE DEAN

TO: Dean Kenneth I. Shine, M.D.
FROM: Norman Cousins

Having just completed ten years at the School of Medicine, I think it may be in order to report to you on my experiences since Dean Mellinkoff invited me to join the faculty in 1978.

During the past decade, my work at UCLA can be summarized under five headings:

1. Lectures to and meetings with medical students.

2. Work with patients who needed to be freed of panic or helplessness, generally at the request of their physicians.

3. Research on the biochemistry of the emotions in conjunction with the work of the UCLA Task Force on Psychoneuroimmunology.

4. Visits to other medical schools, health centers, and hospitals.

5. Memberships on various committees and commissions, such as, Special Medical Advisory Group of the Veterans' Administration Hospitals, the Governor's Cancer Advisory Board (California), Scientific Advisory Board of the Institute for the

Advancement of Health, Special Consultant to the Comprehensive Cancer Center at Duke University and to the Center for Health Communication at Harvard.

Obviously, all these functions were interrelated. The meetings with medical students drew upon my experiences with patients; the meetings with patients drew upon results of research projects, the visits to other medical centers involved patients, students, physicians, etc., etc. In any case, the following is an attempt to provide some details on the various aspects of my work and experiences since coming to the medical school.

In a memo to the faculty some months ago, you pointed out that progress in diagnostic technology now makes it possible for physicians to shift their emphasis to treatment. I am heartened therefore by plans to give increased attention in the curriculum to patient-physician relationships. In that light, I hope the physician's communication techniques can receive major attention. Few things I have learned in the past ten years have hit me more forcibly than the need of the physician to pay attention to the environment of treatment. What stood out most in the meetings I have had with more than five hundred seriously ill patients over the years was that their condition deteriorated sharply at the time of the diagnosis. The moment they had a label to attach to their symptoms they became perceptibly worse. The label had certain connotations. These connotations led to panic, helplessness, and depression, all of which imposed physiological penalties—and all of which underlined the overriding importance of the physician's ability to communicate a diagnosis in a way that challenged but did not crush the patient.

Ten years ago, I thought that iatrogenic problems were represented solely by physician error in the treatment of patients; i.e., mistakes in diagnosis or surgery or harmful effects of prescribed medication, etc. It became apparent to me over the years, however, that errors or failures in communication can be equally iatrogenic. The psychological devastation caused by inartistically delivered diagnoses may be no less serious than mistakes in medication or surgery. The doctor who gives a terminal date cannot be absolutely certain in every case that his prediction is correct—yet that prediction can have the effect of a hex and impair effective treatment. I can understand why physicians fear they may be at risk if something on the downside should occur that comes as a surprise to the patient. It is also true, however, that the patient who is emotionally crippled by the manner of diagnosis is halfway down the hill even before treatment begins.

The emotional state in which the patient leaves the doctor's office can set the stage for optimal treatment or for noncompliance, defeatism, and depression.

Perhaps the single most important fact emerging from my experience and observations since coming to UCLA has to do with the importance of the doctor's role beyond the prescription pad, and, in particular, his ability to invoke the patient's own bodily resources. How the physician employs the art of responsible reassurance, how he identifies and deals with basic causes of illness, how he educates the patient about the essential robustness of the human body—all these questions are not secondary but primary in effective medical procedure.

Why do some diagnosticians create negative expectations that can only do serious damage to the patient? Some of my friends on the medical faculty say the physician is obligated to tell the truth; he has no choice. Others, as mentioned earlier, point to the possibility of malpractice suits if something should happen that the physician hadn't anticipated.

Some physicians, however, like Mitchel Covel and Omar Fareed, will not allow themselves to be pressured into making negative predictions, no matter how extreme the case. They tell the patient that no one knows enough to make a precise forecast, that remissions continue to be reported that were not considered possible, and that the doctor should be concerned not with prophecy but with heroic and compassionate treatment.

As for the argument that it is essential for the patient to get his affairs in order, Dr. Covel tells his patients it is a matter of common prudence even for healthy persons to be tidy about such things. The precariousness of human existence—whether with respect to accidents or to other unexpected events—dictates that every mature person should give attention to such basic matters. Dr. Covel tells his patients that he himself periodically checks on his insurance and his will. Dr. Fareed emphasizes to his cancer patients that so long as the medical journals publish accounts of unexpected remissions, he feels justified in avoiding predictions. His concern is focused on the best possible treatment.

I hope you feel I am acting responsibly when I emphatically urge medical students not to do or say anything that will rob patients of their hopes or deplete their courage. These hopes are the physician's strongest ally. They create a stage on which both the doctor and patient can work together in responding to the challenge of serious illness.

I am profoundly impressed by your new plan for curriculum revision that calls for emphasis on everything involved in physician-patient relationships. None of the courses we *now* give deals *adequately* with the physician's communication skills; none really emphasizes the environment of medical care. None gives sufficient weight to the importance of the physician's style or the way in which the physician's manner can affect the outcome of a case.

I must not unbalance the equation. If I found patients who were disadvantaged by the inartistic communication styles of some physicians, I must also emphasize how impressed I was by the favorable changes in the condition of patients whose physicians understood the importance of creating an environment for effective medical care.

Many of these patients experienced a forward surge as the result of the sensitive way the physician managed the diagnostic session. The physician put his emphasis on challenge rather than verdict. He clearly identified everything that modern medical science had to offer and he just as clearly identified the resources of the patient—physiological, psychological, emotional—that help to create an auspicious environment for treatment. The physician recognized the downward pull of hopelessness and a crushed will to live. While he was careful to avoid creating false hopes, he was equally careful to avoid creating false fears. Since remissions occur that are beyond scientific expectations, the physician didn't want to weaken the possibility of such a remission in the case at hand by causing the patient to slide into the panic-depression cycle that is a medical problem in itself.

Let me move on to the matter of medical education itself. I have tried to speculate on what Dr. Abraham Flexner would say if he were alive today and undertook another survey of the state of American medical education. He would be pleased, I think, by the present emphasis on the scientific method and by the comprehensive scientific content of medical education. But I think he would also observe that education—the process of transmitting knowledge to students—is no longer the main business of medical schools. The physicians on medical faculties, generally speaking, are not hired or promoted according to their demonstrated teaching abilities. They are chosen and esteemed according to their standing as scientists and medical investigators.

There is a tendency, I think, to take medical pedagogical skills for granted. But the qualities that go into the making of a great medical scientist are not necessarily the same as the qualities that go into the making of a skilled teacher. The ability to transmit knowledge is a very special skill; indeed, it is a science in itself. As a product of a school of education, I am perhaps unduly sensitive to the importance of educational techniques; and I must report my impression that education *qua* education may not stand high enough in the mission of medical schools.

In this connection, I am profoundly encouraged by your plan to create a new Center for Health Education, which will address itself to these major problems.

A great deal of attention is now being given to curriculum reexamination and reform, not just at UCLA but at medical schools across the country. I was tremendously heartened during your talk before the psychiatry faculty when you spoke of the need to broaden the education of medical students so that they would understand not just the species of the illness before them but the context of that illness and, in general, the world surrounding the patient. Some physicians have a tendency to put technology ahead of the direct examination of the patient. Indeed, the overreliance on exotic and expensive technology may be one of the major problems confronting modern medicine. A patient who spends more time with diagnostic machines than with the physician may be deprived of one of the main elements of effective treatment—the confidence and reassurance that only a skilled human being can give to another.

In terms of specific measures that could be helpful in creating an optimal environment for medical or surgical treatment, I can think of nothing more constructive or useful than taking the wording of the informed-consent papers away from the lawyers. The lawyers' forms may protect the physician and the hospital, but, in the long run, they are subtractive and even destructive. By informing the patients so graphically and legalistically of all the things that can go wrong, these forms actually create a range of expectations that undermine effective treatment. I believe I told you of my experience in speaking at a national conference of anesthesiologists, where I learned about an alarming episode that occurred at the Tulane General Hospital in New Orleans. There had been a sudden outbreak of heart arrests under anesthesia. An investigation connected the outbreak to two anesthesiologists who would inform the patient of all risks of anesthesia just before they went into surgery. A more sensitive and intelligent way was then devised of providing "informed consent"—and the outbreaks disappeared.

Isn't it possible that our informed-consent papers should be drafted by writers rather than by lawyers? Let the lawyers review the text, but keep them away from drafting documents that can infect and damage the minds of patients, to their detriment and the detriment of the medical center. Wouldn't it be useful to bring in professional writers who could prepare the texts for informed-consent papers that would inform patients without leaving them in a state of panicky anticipations? If our aim is to optimize treatment, we certainly don't want to cripple the patient psychologically even before the treatment begins, which is the unfortunate effect of many of the informed-consent documents.

Our Task Force in Psychoneuroimmunology not only undertook and sponsored relevant research at UCLA but supported similar research at other medical centers in this country and abroad.

As you know, we held a world conference on psychoneuroimmunology at UCLA's conference facility at Lake Arrowhead in March 1988. The papers delivered at this conference sought to meet the doubts of those who asked for hard evidence that the workings of the mind can actually produce physiological effects. They wanted to know what pathways and mechanisms were involved. You yourself called for definitive scientific data that could show that states of mind had biological effects. You would have been impressed, I think, with the depth and range of the presentations on this point. Drs. Ronald Glaser and Janice

Kiecolt-Glaser reported a series of controlled experiments involving a significant number of medical students. The experiments demonstrated a clear relationship between nervousness and apprehension as examination time approached and increased susceptibility to a wide range of illnesses. This connection was verified by tests showing a lessening of immune capacity during the period of increased stress.

Dr. Neal E. Miller, of The Rockefeller University, and Yale University, reported at Arrowhead that the adverse psychological effects of stress can be reduced by certain coping strategies, including positive emotions, social support, and physician reassurance. Dr. Miller also cited evidence showing adverse effects on cardiac and gastrointestinal function as the result of increased tension and emotional wear and tear.

Also at Arrowhead, Dr. Branislav D. Janković, of the University of Belgrade, Yugoslavia, who published the first paper on the importance of the thymus in immunity and who was among the first to show the effect of brain lesions on immune responses, reported his recent work in linking immune activity to pathological brain function and behavior.

He presented results from his laboratory work showing that anti-brain antibodies (immune cells that attack brain cells) cause abnormal brain function and behavior. He has correspondingly shown that individuals with neurological and psychiatric disorders such as dementia, schizophrenia, depression, alcoholism, and Parkinson's disease display high levels of immune reaction to injections of brain tissue. This research corroborates the connection between diseases involving the nervous system and the immune system.

Other laboratory studies linking behavior to immunity were reported. Dr. Robert Ader of the University of Rochester reviewed his own pioneering efforts in the conditioning of the immune system to respond to placebos. Even though conditioned suppression of the immune system may be slight, it can be used effectively to decrease the drug dosage required to lower the mortality of rats prone to a lupuslike autoimmune disease.

Dr. Novera Herbert Spector, of the University of Alabama, Birmingham, and Georgetown University Medical Center, also reported a research project involving mice in which he and his colleagues demonstrated that immune conditioning resulted in enhanced immunity, delay or disappearance of tumor growth, and increased survival.

Dr. Mark L. Laudenslager of the University of Colorado, who previously demonstrated that animals that learned to control stress (in contrast to "helpless" animals) registered no ill effects on immunity, reviewed his research on the effect of social and environmental conditions on the emotional and immunological status of nonhuman primates. He cited numerous examples of the complexity of the associations between behavior and immune response. For example, he found that animals experiencing separation stress showed diminished lymphocyte stimulability, while exhibiting increased natural-killer-cell activity. He has also found that the nature of the species, the precise circumstances of the environment, and the hierarchical status of the animal seem to be significant factors in measures of immunity.

Dr. Rudy E. Ballieux of the University of Utrecht, the Netherlands, emphasized the complexity of the interaction between behavior and immunity. He reviewed his own experiment with rats, in which both passive behavior and suppression of antibody responses were directly proportional to the intensity of electroshock administered. However, he also demonstrated that the application of a specific neuropeptide to the brain, while it could extend the passive behavior, nonetheless arrested the immune suppression. The apparent contradiction, he said, has not yielded to plausible explanation.

Dr. Elena A. Korneva of the Institute for Experimental Medicine in Leningrad, USSR, spoke about Dr. Ballieux's reference to the fact that antibody suppression was not increased when neuropeptides were applied to the brain. Her own research, she said, provided another example of the apparent contradiction that certain kinds of stress enhance antibody responses, whereas drastic or prolonged stress appears to inhibit immunity.

One way or the other, the evidence seemed to be clear that behavior has a verifiable effect on immunity. Dr. Miller pointed out that just the presence of the physician, especially when he enjoys the strong confidence of the patient, can have beneficial physiological effects. The same is true, he said, if the patient's own mind-set is one of hopefulness rather than helplessness. Under negative circumstances, he said, the downside effects can be observed not only in terms of immune impairment but on the body's other systems—cardiovascular and gastrointestinal included.

The intricacies of psychoneuroimmunologic phenomena were further detailed by a number of presentations on mediating mechanisms (a biological circuitry that explains observed connections between psychological states or behavior, and immunity). These reports were of special interest because the fundamental premise of psychoneuroimmunology is that the brain, the endocrine system, and the immune system are in constant states of interaction with each other.

Dr. David L. Felten, associated with Dr. Ader of the University of Rochester, provided a detailed account of the pathways involved in the nervous system stimulation of immune cells. Norepinephrine, released by nerve cells, binds to receptors on immune cells. These findings support the theory of neurotransmission to immune cells. He also reported that drug-induced suppression of norepinephrine lowers some immune responses (e.g., T cells), but enhances B-cell and NK-cell activity, suggesting multiple roles for norepinephrine in the regulation of the immune system.

Felten's references would have been of great interest, I think, to Walter Cannon, who, more than a half-century ago, was able to demonstrate an increase of 10 to 15 percent in the population of red blood cells as the result of increased activity by the spleen when touched off by heightened emotional states.

Dr. Karen Bulloch, of the University of California at San Diego, who gained attention for her work on the innervation of the thymus gland, showed how specific immune cells possess receptor sites that enable them to connect with a variety of neurotransmitters. She presented evidence that the nervous system plays a critical role in the development of the immune system.

Dr. Walter Pierpaoli, of the Institute for Biomedical Research in Quartino-Magadino, Switzerland, reported findings of his research (in collaboration with Dr. Georges J. M. Maestroni of the Laboratory for Experimental Pathology in Locarno, Switzerland) that the neuroendocrine system is involved not only in the development of but in the *ongoing* regulation of the immune system. For example, interference with the cyclical release of the hormone melatonin (released by the pineal gland) profoundly handicaps immunity. Pierpaoli's research has also shown that the immune system has important regulatory effects on the neuroendocrine system.

Dr. Korneva, whose work has been described more fully in Chapter 5, also presented evidence from her laboratory that varying levels of immune responses produce correspondingly measurable changes in chemical and physiological activity within the hypothalamus.

Dr. Hugo O. Besedovsky, of the Swiss Research Institute in Davos-Platz, Switzerland, detailed a feedback loop between the immune system and the brain that he hypothesized plays a role in preventing overactivity of the immune response (such as autoimmunity). For example, activated monocytes and macrophages produce interleukin-1, which in turn increases corticotropin-releasing factor activity in the hypothalamus, which results in an increase in adrenocorticotropic hormone and corticosterone blood levels, and decreased immune activity.

Dr. Ballieux also reported that his research, as well as the research of others, has established the fact that immune cells release neuropeptides and neurohormones that are indistinguishable from those released by the pituitary gland. Ballieux raised the possibility that the neuropeptides and neurohormones have parts to play in the self-regulation of the immune system.

Dr. Candace Pert, referred to earlier, offered an integrative perspective based on her own work in the field of neuropeptides, which serve communicating functions that go on constantly between the brain and the body's various systems. This communication is by no means to be regarded as a one-way process from brain to body, but rather as a multiple process in which the brain receives and processes messages from the various parts of the body. Because neuropeptides are capable of binding to receptor sites on immune cells and brain cells, Dr. Pert deduced that the human immunodeficiency virus that causes AIDS attaches to common receptors in the brain as well as in the immune system.

Such a phenomenon would possibly explain the fact that AIDS patients often experience degenerative cognitive function and depression. Dr. Pert subsequently isolated a compound capable of being attached to common neuropeptide receptors—

blocking the AIDS virus from infecting both immune cells and brain tissue.

UCLA has reason to be proud, I think, of the presentations at Lake Arrowhead by members of our Task Force. Dr. Herbert Weiner, whose book *Psychobiology* anticipated in 1976 many of the prime developments in mind-body studies, proposed that health and disease involve breakdowns of communication (within and between cells) leading to abnormal regulation of bodily functions and behaviors—a theory that Dr. Felten remarked could unify the life sciences. Dr. John Fahey presented a panoramic view of recent research on the immune system, with emphasis on the multiple interactive forces affecting its workings. He stressed the need to recognize that we aren't dealing with a single coherent system when we speak of the immune system, but a series of subsystems, each of which has specific characteristics that affect the total functioning of the immune process.

Dr. Fawzy Fawzy called attention to the research project in which he and I have been engaged for more than a year. There was significant evidence to show that the liberation of cancer patients from depression can actually increase the wide array of natural killer cells, thus activating the anticancer capability of the immune system. Techniques for relieving the depression of malignant melanoma patients included participation in support groups as well as attitudinal and behavioral therapy. What was most heartening to me about this research project was the clear evidence of improvement in the clinical condition of the patients. Compared to the control group, the research patients as a whole showed greater evidence of enhanced immune activity. They demonstrated the importance of emotional and social factors in optimal treatment.

This is a rough summary of what happened at Lake Arrowhead. What was most significant about the meeting, of course, was that the participants accepted your challenge to deal not just with their perceptions of cause-and-effect in psychoneuroimmunology but to apply the strictest standards of scientific inquiry to their work and in their reports. The quality of the research that underlay and buttressed most of the Arrowhead reports was most impressive.

I do, however, perceive a problem. Physicians who doubt

that physiological effects can proceed out of psychological causes have been most explicit in expressing their reservations. Those reservations are to be taken seriously. What do we do to persuade these doubters to examine the very evidence they have called for? Equally important: How do we go about infusing knowledge developed by such research into the medical curriculum?

Some physicians criticize our ideas in the name of science, but they see nothing unscientific in arbitrary criticisms that ignore or bypass serious research. Hence the question: How do we get the doubters and the critics to scrutinize the results of the research they ask for, and then, if they still have their doubts, how do we get them to specify in what respects they think the research is faulty or inadequate? How do we initiate a serious dialogue based on fact rather than random impressions?

The Task Force has undertaken or sponsored at least two dozen studies on the effects of psychological and psychosocial factors on the immune system. These studies were pursued at various medical centers across the country and tend both to reinforce and to replicate one another.

Now that I am embarking on my second decade at UCLA, I want to thank you for your generous support of my work at the medical school. I recognize how unusual it is for a layman to come into the medical community, especially into a field as new and complicated as psychoneuroimmunology. But you and your predecessor, Sherman Mellinkoff, have provided both guidance and encouragement. I am similarly grateful to Jolly West and Carmine Clemente for their special efforts on my behalf.

I had no idea when I joined the medical faculty in 1978 that the next decade would find me so completely caught up in the world of physicians, patients, and medical research. But it has been a productive and rewarding ten years and I have no regrets about the decision to embark on a new path.

The obsession that brought me to the medical school—the obsession to understand the connection between the positive emotions and bodily changes—now gives way to a new and perhaps even larger obsession. The subject of physician communication needs to be raised to the same level of importance in medical education as pathology, physiology, biochemistry, anatomy, or any of the other subjects that are considered basic in the training of physicians.

•

Let me now summarize the major conclusions emerging from ten years among the Aesculapians:

1. The human body is far more robust than people have been led to believe. Public education in health matters has tended to make people overestimate their weaknesses and underestimate their strengths. The result is that we are in danger of becoming a nation of weaklings and hypochondriacs. Franz Ingelfinger's estimate that 85 percent of all illnesses are self-limiting should serve as the cardinal fact in the reeducation of the American people, just as it should dictate a restrained method of treatment, especially with respect to medications.

2. Patients have a tendency to become panicky. The physician's ability to reassure the patient is a major factor in activating the body's own healing system. Reassurance can help to optimize treatment.

3. The wise physician gives careful attention to the environment of medical care. Circumstances surrounding treatment can affect the result.

4. A strong will to live, along with the other positive emotions—faith, love, purpose, determination, humor—are biochemical realities that can affect the environment of medical care. The positive emotions are no less a physiological factor on the upside than are the negative emotions on the downside.

5. Depression is a demonstrated cause of physical ill health, including deleterious effects on the immune system. Equally striking is the fact that liberation from depression produces an almost automatic boost in the number of disease-fighting immune cells. The best depression-blockers are a strong will to live, blazing determination, and a sense of purpose best expressed perhaps by interesting, useful activity. The physician's role in attending to the psychological needs of patients cannot be overemphasized.

6. Even the most positive attitudes by the patient are no guarantee of a cure, but they can help create an environment conducive to medical care and can enable both patient and physician to get the most out of whatever may be possible.

7. Patients tend to move along the path of their expectations, whether on the upside or on the downside.

8. So long as unexpected remissions occur—and the medical journals are the best places to find such ongoing evidence—both doctors and patients are justified in hoping for the best and working for the best.

9. Challenge creates a better environment for treatment than does a grim verdict. If the physician intends to treat a serious disease, he must convince the patient both to make a special effort and to believe that the effort is worthwhile.

10. Family, friends, support groups—all can be helpful in combating the emotional devastation that frequently follows a serious diagnosis. Just as the brain tends to convert bad news into panic and helplessness, so strong support from family and friends can help maintain or restore emotional equilibrium.

11. Medical technology is not the ultimate arbiter. The mood in which a patient takes a diagnostic test can affect the outcome of the test. This is especially true of cardiac testing. The circumstances of the test, the ambience of the testing room, and the mood of the patient can sometimes affect the results. In any case, modern technology is not a total replacement of the physician's diagnostic skills.

12. The ability of the physician to listen to the patient is no less important than the printouts of diagnostic technology. An understanding of all the factors leading up to an illness is no less important than the identification of pathological organisms.

13. Finally, the past ten years have given me an enlarged respect for the men and women who have committed themselves to the practice of medicine. True, I have been troubled by some aspects of medical education and practice, but I would be even more troubled if these observations were to diminish the most important conclusion emerging for me from a decade at the medical school. The large majority of people involved in medicine—students, teachers, nurses, doctors—can hold their heads high in terms of the quality of their work, their connection to a sustained high purpose, and their understanding of the philosophical dimensions of their profession.

It is difficult to say which is the most important of these lessons. If I had the power to bring about only one change in medical education, it would be to double and redouble the attention given to the patient-physician relationship. The way the doctor delivers a diagnosis; the way he creates an environment conducive to effective medical treatment; the way he inspires a patient to become a full partner in a strategy for recovery; the way he takes into account the emotional needs both of the patient and of members of the family—all these factors are involved in an effective patient-physician relationship.

The major advances in modern medical science give substance to the principle that the mind of the patient creates the ambience of treatment. Belief becomes biology. The head comes first.

23

A PORTFOLIO OF RELATED MATTERS

HONOR ROLL

Knowledge of the role of the mind in health and healing goes back to the beginnings of medicine and literature. The Honor Roll, past and present, of those whose work and writings embrace the concept of interaction between mind and body, would include the following but is surely not confined to it. This list, therefore, does not profess to be complete. The following names, in rough chronological order, are an addendum to those whose work has already been described in previous chapters.

HIPPOCRATES (460 B.C.), physician of Greece, was characterized by Plato as "a professional trainer of medical students." A systematic observer and organizer of medical phenomena, he taught his students to observe the life circumstances and emotional states of patients and to earn their confidence. He was opposed to any treatment that might interfere with the natural healing process (*vis medicatrix naturae*) or that might do harm (*primum non nocere,* "first do no harm").

ARISTOTLE (384–322 B.C.), philosopher and biologist of Greece, believed that the soul was housed in and inseparable from the body and that all parts of the body are functionally related to serve the whole. He defined the functions of blood and the various bodily organs, identified the specific locations from which the emotions emanated, and referred to "the affections of the soul" as being "ideas expressed in matter." His observations of living things laid the foundation for comparative anatomy and embryology.

GALEN (130 A.D.), Greek physician of Rome, advocate of the *art* of medicine—meaning the need for sensitivity and common sense in treating patients. He emphasized the role of the patient in any strategy for recovery. What was the patient's own view of the illness? What kind of remedial treatment did the patient select for himself? He also observed the connection between melancholy and malignancy and classified the "passions" (emotions) as a non-natural cause and also as a factor in treatment.

MOSES MAIMONIDES (1135–1204), Jewish philosopher and physician who was one of the earliest advocates of preventive medicine and urged restraint in the use of drastic techniques such as surgery. For example, he emphasized the health benefits of proper exercise, nutrition, rest, and climate—adjusted to an individual's own bodily constitution. He also explored the bodily manifestations of the emotions, the importance of which was underscored by his recommendation that the ideal type of exercise was one that brings the emotions of happiness to the soul—emotions that he maintained would often, in themselves, complete the healing process. He brought a strong philosophical dimension to the treatment of human beings.

PARACELSUS (1493–1541), itinerant German physician and alchemist, sometimes called the father of chemistry and the reformer of materia medica, acknowledged factors beyond the control of the physician in healing (for example, the human spirit and the healing power of nature). He declared that body and mind were indivisible and observed the role of emotions in mental and physical diseases. He urged doctors to develop their powers of observation instead of relying so heavily on books and standard practices of the day, which at that time involved the measurement of "humors," or bodily secretions.

BENEDICTUS DE SPINOZA (1632–1677), Dutch-Jewish philosopher, and foremost exponent of seventeenth-century Rationalism, rejected the notion, proposed by René Descartes in the same century, that the mind and body were two distinct entities. He postulated instead a single, infinite substance, which he called God, or Nature, of which the mental and material were experiential components. Therefore, to every mental event there was a corresponding physical event, and vice versa. He described the emotions and identified the occasions that elicited them, believing that contradictory emotions either produced fluctuations or neutralized each other.

OLIVER WENDELL HOLMES (1809–1894), writer, philosopher, physician, believed in the power of the physician's attitude in the recovery of a patient. He was chastised by his colleagues for having accused medical practitioners of needless overprescribing of drugs. He asserted that, with the lessening of emphasis on pharmacologic treatments, more attention would be paid to the hygiene, comfort, nourishment, and general care of the patient. In this regard, he underscored the importance of thorough training for all persons involved in the direct care of patients.

In a lecture for the one-hundredth anniversary of the founding of Harvard Medical School, he illustrated the value of careful attention to detail: " 'Will you have an orange or a fig?' said Dr. James Jackson to a fine little boy. . . .'A fig,' answered Master Theodore with alacrity. 'No fever there,' said the good doctor, 'or he would certainly have said an orange.' " He believed that the common instincts of humanity would inevitably produce a balance between respect for and overreliance on medical science.

CLAUDE BERNARD (1813–1878), French physiologist, was instrumental in establishing principles of scientific experimentation in the field of physiology. In discovering that the brain was involved in regulating blood sugar levels, he developed the concept that the body's internal environment was governed by the law of stability—which ultimately contributed to what Walter Cannon called homeostasis, or the self-regulation of vital processes.

JEAN MARTIN CHARCOT (1825–1893), French anatomical pathologist and teacher of Sigmund Freud, first described "conversion hysteria"—ways in which emotional disturbances became physical symptoms, sometimes disabling. He looked beyond the flaws in the "animal magnetism" notions of Franz

Anton Mesmer to the possibilities of hypnosis and the powers of suggestion.

WILLIAM JAMES (1842–1910), philosopher and psychologist of Harvard University, led the philosophical movement of pragmatism and the psychological movement of functionalism (or the application of biological disciplines to mental science—treating thinking and knowledge as instruments in the struggle for survival). He outlined a myriad of mental processes (including emotions, memory, reason, time perception, stream of consciousness, etc.) demonstrating the complexity of the human mind. He acknowledged the existence of powerful and accessible reservoirs of conscious energy that could be tapped for healing, and suggested that human beings tend to live too far within their potentialities.

SIR WILLIAM OSLER (1849–1919), Canadian physician of Johns Hopkins University Medical School and Oxford University in England, was considered the most eminent clinician of his time. He was a powerful balance wheel in medicine, reconciling stern scientific and technological aspects of medicine with more abstract factors such as the art of medicine. He called attention to the role of the emotions in the "functional diseases of the nervous system," including neurasthenia, joint pains, tremors, spasms and tics, migraines, and some forms of paralysis. When praised for his success as a physician in curing organic diseases, he was inclined to shift the credit to the patient's faith in the treatment and to the quality of nursing care.

IVAN PETROVICH PAVLOV (1849–1936), Russian physiologist, who demonstrated that salivary and gastric secretions could be elicited by sounds associated with food—providing evidence of a connection between expectations and nervous system activity. This research led to other studies showing that many bodily systems (including the endocrine and immune systems) could be conditioned. He also developed physiological measures of emotional states and nervous system activity.

WILLIAM HENRY WELCH (1850–1934), pathologist and first dean of the Johns Hopkins University Medical School, was regarded by many as a prime mover in the development of modern medical practice and education in this country. He developed a curriculum that demanded of its students both a rigorous study of physical sciences, and an active involvement in clinical and laboratory work. He recognized the importance of patient attitudes and moods in the alleviation of symptoms.

WALTER BRADFORD CANNON (1871–1945), neurologist and physiologist of Harvard Medical School, studied the role of the emotions in fostering a wide range of bodily changes as well as the electrochemical channels by which specific emotions are expressed. His pioneering studies on the role and functions of the adrenal gland led to the articulation of his "fight or flight" hypothesis. He was an effective bridge between medical science and the public, gaining wide recognition for such books as *The Wisdom of the Body*, in which he introduced people to the vast and wondrous universe that existed within themselves. He observed and used the term *homeostasis* to describe the body's natural tendency to maintain its vital balances.

LAWRENCE J. HENDERSON (1878–1942), biochemist of Harvard Medical School, discovered the chemical means by which acid-base equilibriums are maintained in bodily fluids. His observation of chemical interactions led him to believe that chemical evolution, the creation of life, and biological evolution, proceed logically and adaptively (i.e., by design) and not at all accidentally. Henderson's belief in the order of nature led him to conclude that the function of ideas and consciousness is to modify the course of mechanical processes and, therefore, that the organism is not merely a physical structure, but a "psychophysical" whole.

FRANZ GABRIEL ALEXANDER (1891–1964), physician and psychoanalyst of the University of Illinois Medical School, sometimes referred to as the father of psychosomatic medicine, played a leading role in identifying specific physical disorders involving changes in bodily function (disorders including peptic duodenal ulcer, bronchial asthma, essential hypertension, rheumatoid arthritis, and ulcerative colitis), initiated by particular constellations of unconscious conflicts developed in early life and personality traits emanating therefrom. He contributed to the belief that strong emotions triggered by unconscious conflict-activating situations, combined with physical predispositions, can result in autonomic, hormonal, and other reactions to produce actual bodily changes.

KARL J. MENNINGER (1893–), psychiatrist and cofounder of the Menninger Foundation in Topeka, Kansas, has articulated that psychiatry ought to concern itself less with how to label a syndrome and more with the pragmatic and dynamic goal of what to do about it. He has delineated internal and external factors involved in the forces of recovery which include attitudinal subtleties on the part of the physician and intangible positive emotions such as love, faith, and hope, which he termed "sublime expressions of the life instinct." Dr. Menninger also outlined the process by which a thorough and skillful diagnosis aimed at understanding the patient in relationship to the total environment could lead to a treatment geared toward developing new insights and potentialities in the patient. He has observed such treatments to produce optimum well-being or transcendence of illness in his patients that exceed standard expectations for "recovery."

HAROLD G. WOLFF (1898–1962), neurologist of Cornell University Medical College in New York, demonstrated the role of life situations and emotional turmoil in altering the function of nearly all bodily systems. He found that when his patients discussed personally meaningful topics, their emotional responses produced short-term changes in bodily functions. He showed that the ability of the body to adapt to circumstances or to protect itself against a harmful agent was related to the patient's own experience or interpretation of the threat. He cited the failure to adapt as a cause of disease.

RENÉ JULES DUBOS (1901–1982), philosopher, microbiologist, and environmentalist of the Rockefeller Institute for Medical Research in New York City, and editor of the *Journal of Experimental Medicine,* isolated antibacterial substances from certain soil microorganisms, thus playing a major part in the development of antibiotics. His writings have stimulated studies of the way in which human beings and animals respond and adapt to environmental challenges. He was no less concerned with the health of human society as a whole than with individual health. Toward the end of his life, he wrote books that reflected his concern about the life support system of the planet because of almost total disregard of the poisons being spewed out by industrial smokestacks, incinerators, and combustion engines. He regarded human life as a totality with stern requirements.

HELEN FLANDERS DUNBAR (1902–1959), psychiatrist of Columbia University in New York and first editor of the journal

Psychosomatic Medicine, was instrumental in defining this field as a psychological approach to medicine in which mind and body were an entity. She helped to launch the field by pulling together fragments of knowledge pertaining to psychosomatic medicine in her landmark book, *Emotions and Bodily Changes*. She found that patients afflicted with specific diseases were characterized by relatively consistent behavioral characteristics.

HENRY K. BEECHER (1904–1976), anesthesiologist and ethicist of Harvard Medical School, documented the power of the placebo in producing a wide range of bodily effects—both beneficial and toxic—in a large number of disorders ranging from autoimmune disorders such as rheumatoid arthritis to the common cold. He also reported that the greater the need for help, the greater the impact of the placebo.

HANS SELYE (1907–1982), neuroendocrinologist of the University of Montreal, Canada, did more, perhaps, than any physician of his time to inject the concept of stress into the public consciousness. His books on the subject served to connect the "wear and tear" of life's circumstances to specific disease. He observed general physiological and hormonal patterns of change under varying stressful conditions. The term he used to describe the process was *general adaptation syndrome* (GAS), or the "stress response." The GAS consisted of three phases—an "alarm" reaction phase during which the body mobilizes its defenses, followed by a stage of resistance, and finally culminating in exhaustion and death. He challenged the scientific community to develop ways of combating disease by fortifying the body's own defense mechanisms against distress.

JEROME D. FRANK (1909–), psychiatrist of Johns Hopkins University Medical School, observed that diverse modes of medical treatment owed their success or failure to the patient's state of mind and expectations and not solely to the treatment regimen itself. He has also maintained that the physician's therapeutic power could be enhanced by fostering "the faith that heals" and by nurturing a humane environment. He has regarded any treatment of an illness as grossly deficient if it does not also minister to the human spirit.

NEAL E. MILLER (1909–), physiological psychologist of The Rockefeller University, New York, and Yale University, demonstrated that involuntary bodily responses regulated by the autonomic nervous system could be elicited or controlled by rewarding (or reinforcing) the behaviors that produce those physiological reactions. This showed that some measure of control was possible over autonomic functions. He has been a major figure in the development of behavioral medicine—a field in which thoughts or actions interact with physiological processes. In biofeedback techniques, for example, individuals are able to lower their blood pressure, relax muscles, or control their bowels. Physical or mental exercises used to bring about relaxation or other health-related behavior change are other key examples of treatment modalities in behavioral medicine.

FRANZ J. INGELFINGER (1910–1980), gastroenterologist of Harvard Medical School and editor of the *New England Journal of Medicine*, challenged both physicians and medical critics to confront problems in the quality of medical care. He urged physicians to recognize the importance both of providing the patient with reassurance and exercising restraint in the use of exotic medications. He suggested that moderation in the use of medical jargon with patients might improve communication. He saw sensitivity by the physician to the emotional needs of the patient as inseparable from efficient medical practice. He had a good sense of humor, suggesting that some of the problems of medical education might be solved by requiring previous hospitalization as a requirement for entry into medical school.

GEORGE L. ENGEL (1913–), psychiatrist of the University of Rochester School of Medicine and Dentistry, articulated the "biopsychosocial" view of health—the view that takes into account the interaction of biological, psychological, and social (environmental) factors in the onset of physical disorders. For example, he identified the importance of life events that elicit feelings of loss to an individual, as a result of his or her constitutional and developmental background, in triggering disease processes. He advocates the intensive study of individual patients as a source of new information about factors involved in health and disease and encourages physicians to heed their intuitions and observations, emphasizing the value of listening to and interpreting patients' theories about their conditions.

JOEL ELKES (1913–), psychiatrist of The Johns Hopkins University, the University of Louisville, Kentucky, and McMaster University, Canada, was one of the founders of the field of psychopharmacology (the study of the action of drugs on the mind) and a leading proponent of the medical humanities. He developed advanced programs in behavioral medicine for medical students and has been involved in efforts to establish the arts as an accepted therapeutic modality.

VERNON S. RILEY (1914–1982), of the Pacific Northwest Research Foundation and the Fred Hutchinson Cancer Research Center, Seattle, Washington, conducted pioneering research indicating that emotional distress contributes to cancer growth through alterations in hormones and immunity. He also contributed much to the precision of psychoneuroimmunologic research—showing that such factors as intensity, duration, and timing of stress; age; sex; and early life experience have different effects on immunity.

JONAS EDWARD SALK (1914–), fellow and founding director of the Salk Institute of Biological Studies in San Diego, California, is known primarily for his work in developing a vaccine for poliomyelitis. Observing a resemblance between cells of the central nervous system and cells of the immune system, he speculated that disease may involve genetic, behavioral, nervous system, and immune system interrelationships, and that the pattern of development of the nervous system may be similar to that of the immune system. These ideas give substance to the underlying tenets of psychoneuroimmunology.

STEWART WOLF (1914–), physician of the University of Oklahoma, carried out advanced placebo research showing that certain medical procedures or "bogus" pills can trigger significant beneficial or detrimental effects—effects sometimes strong enough to override the effects of an actual potent drug. He also observed that placebo effects can be especially effective when the doctor himself is unaware that the "drug" being administered is actually a placebo. These observations led Wolf to conclude that all bodily organs and systems are susceptible to such influence; in short, that the attitude of a physician alone may produce significant beneficial or adverse effects on a patient's health and well-being that are totally separate from any medication administered.

AARON FREDERICK RASMUSSEN, JR. (1915–1984), immunologist and associate dean of the UCLA School of Medicine, explored the relationship between stress and susceptibility to viral infection. He found that stress diminished the size of immune organs and numbers of immune cells and increased the severity of illness reactions to viral exposure.

DAVID M. KISSEN (1916–1968), thoracic surgeon and director of the Psychosomatic Research Unit of Southern General Hospital in Glasgow, Scotland, was a major figure in the study of psychosocial factors and cancer incidence. He observed in a series of well-designed studies a number of patterns characteristic of cancer patients, among them: childhood trauma experienced in parental relationships, long-term difficulties at home or at work, and, most important, poor outlets for emotional expression. His research suggested that an individual's emotional response to a life event was more critical than an event itself in the genesis of cancer.

KARL H. PRIBRAM (1919–), neurophysiologist of Stanford University, has been studying neurophysiology in the context of psychological theory and direct observations of human behavior, and has applied the principles of philosophy, such as William James's pragmatism, to neurophysiological research (e.g., that evolutions in theoretical understanding require acceptance of the fact that current understandings of science are but approximations of an infinite realm of possibilities). His approach to the study of the brain enabled him to infer complex neurophysiological pathways of learning and memory based on observations of human perception, motivation, expectation, emotion, and behavior.

LAWRENCE LeSHAN (1920–), research psychologist of the Institute of Applied Biology in New York, conducted extensive pioneering work regarding the cancer-prone personality that led him to identify several psychological characteristics that seemed to typify cancer patients (including such factors as the inability to express aggression and disruption of a parental relationship in early childhood). He concluded that personality factors have some bearing on the observed association between traumatic life events (most notably, the loss of a significant emotional relationship) and the development of cancer, and he speculated that specific psychological attributes could be linked to particular types and locations of cancer.

JIMMIE C. D. HOLLAND (1928–), psychiatrist of the Memorial Sloan-Kettering Cancer Center, New York, has been evaluating the psychosocial needs of cancer and AIDS patients and has called attention to the need for emotional support in such cases. She has been utilizing a variety of psychological techniques, ranging from relaxation training to talking with other patients to alleviate their feelings of confusion, anxiety, isolation, helplessness, and defeat; and to control nausea associated with chemotherapy. She has found that the length of cancer survival can be affected by the degree of understanding of treatment and subsequent compliance. Her work underscores the importance of the physician-patient relationship.

ONGOING STUDIES IN BIOCHEMICAL PATHWAYS

The main thrust of the Task Force was of course to sponsor research in mind-body reactions, with special emphasis on the biochemistry of the emotions. Here is a brief description of that research which may be of interest to readers.

Dr. Faisal Abdel-Kariem, visiting scholar in the Department of Microbiology and Immunology at the UCLA School of Medicine, has been examining the effect of a particular family of neuropeptides (including the body-produced painkiller, beta-endorphin) on different cytotoxic cells and lymphokines (e.g. interleukin-2) which stimulate the events of killer-cell activity.

Dr. Francesco Chiappelli, postdoctoral fellow in the UCLA Program in Psychoneuroimmunology, has been observing how the neuropeptide beta-endorphin either directly or indirectly alters NK-cell activity. He has also been investigating the reasons why a substance that mimics cortisol (a hormone produced in response to stress) diminishes lymphocyte activity when administered to normal individuals, but has no effect on the lymphocyte activity of anorexia nervosa patients, whose hormonal disturbances are reflected in elevated levels of cortisol.

Dr. William R. Clark, professor in the Department of Immunology at UCLA, has been evaluating the effect of various endorphins and enkephalins on cytotoxic T cells. Thus far he has found that these particular neuropeptides do not influence the killing activity of cytotoxic T cells. In a step-by-step process of unveiling the effects of neuropeptides on cytotoxic T cells, Dr. Clark plans to investigate whether other neuropeptides affect the killing activity of cytotoxic T cells and whether neuropeptides affect other cytotoxic T cell functions such as the release of lymphokines (immune-regulating chemicals). This study, in concert with the research of Drs. Abdel-Kariem and Francesco Chiappelli and the work of Drs. Jean Merrill and Fergus Shanahan, will clarify the specific ways in which neuropeptides might contribute to the observed influence of the nervous system on the immune system, and vice versa.

Dr. Roger A. Gorski, professor and chairman of the Department of Anatomy at the UCLA School of Medicine and director of the Laboratory of Neuroendocrinology at the UCLA Brain Research Institute, was one of the first medical researchers to discover anatomical differences in the brains of males and females. He has specifically found differences within the hypothalamus—a major center of activity for the regulation of the emotions, neuroendocrine system, and immune system. There is evidence from laboratory studies that neuroendocrine or brain (hypothalamus) dysfunction during development can result in abnormal sexual differentiation of the brain, which can alter reproductive function and sexual behavior. It has also been demonstrated that dysfunction of the thymus (a major organ of the immune system) during development can also result in the same phenomenon. Dr. Gorski hopes to broaden the understanding of nervous, endocrine, and immune system interactions by demonstrating that the immune system is essential for normal nervous and endocrine system development—a notion that has been supported by Drs. Walter Pierpaoli, Hugo Besedovsky, and Karen Bulloch, whose work is discussed elsewhere in this book.

Dr. Gorski has been investigating the possibility that the immune system undergoes a hormone-dependent sexual differentiation. Specifically, he wants to observe how immune activity develops or "differentiates" in response to hormonal exposure during early and later development, and then plans to see whether observed sex differences in immune function can be reversed by extended hormonal treatments during early development. This study may help to explain the marked differences in immune responses between men and women, such as differences in immune response to the emotional distress of bereavement. Men and women are also susceptible to different autoimmune disorders. For example, rheumatoid arthritis is significantly more common in women than in men, whereas ankylosing spondylitis is more common in men than in women. Finally, this research may explain why fetal exposure to the synthetic hormone diethylstilbestrol (DES) increases susceptibility to immune-related disorders—a topic being simultaneously pursued by Dr. Melissa Hines.

Dr. Hines, assistant research biobehavioral scientist in the Department of Psychiatry and Biobehavioral Sciences at the UCLA School of Medicine, is comparing the immunological status and health histories of DES-exposed women with those of non–DES-exposed women to see if compromised immunity plays a part in susceptibility to immune-related diseases. In addition, she is measuring psychological states in order to determine whether immune defects are more apparent during periods of distress. Millions of pregnant women in the United States took DES in the 1940s, 1950s, and 1960s. It has since been found that DES-exposed offspring show an increased risk of developing vaginal and cervical malignancies, autoimmune diseases, asthma, and frequent colds and influenza. Dr. Hines is also exploring whether immune abnormalities stem from permanent emotional differences caused by DES exposure. This study, combined with Dr. Gorski's research on the effects of early hormonal exposure on immune system development, is expected to shed light on the role of the endocrine system in the development of immune organs, and on the causes of and treatment for immune-related diseases.

Dr. Karen Bulloch, director of the Program in Neuroimmune Physiology at the University of California at San Diego School of Medicine, has been developing laboratory techniques for identifying the presence of neurotransmitter receptors on human lymphocytes and has been observing how such receptors affect immune activity. Because she has found that mature immune cells bear greater numbers of such receptors than immature cells, she hypothesizes that the nervous system plays a role in the development of the immune system. She is studying the possibility of regulating immune cell function by altering neurotransmitter activity. She also plans to examine the lymphocytes of individuals with neuroimmune diseases to determine whether or not defects in the development or function of neurotransmitter receptors contribute to immunological abnormalities.

Her work will expand knowledge of the role of the nervous system on immune functioning at a cellular level and may help address questions such as those raised by Dr. Arthur Kling, professor in residence of the UCLA Department of Psychiatry and Biobehavorial Sciences, about the possible role of neurotransmitter disorders in abnormal immunity and behavior (as in schizophrenia).

Dr. Tom Newton, senior resident in the UCLA Department of Psychiatry and Biobehavioral Sciences, and Dr. Jacob Zighelboim, associate professor in the Department of Microbiology and Immunology at the UCLA School of Medicine, have been carrying out research that they hope will enlarge our understanding of the connections between emotions, immunity, and health. They have been studying immune and nervous system interrelationships and the underlying causes of Chronic Fatigue Syndrome. Disturbances in sleep stages have been associated with depression and fatigue and possibly diminished immune system regeneration. Lymphokines—immune regulators such as interleukins—may be involved in sleep regulation, and treatment with these substances produces symptoms akin to Chronic Fatigue. The investigators hypothesize that there will be parallel disruptions in sleep stages, lymphokine levels, and NK-cell activity in Chronic Fatigue Syndrome patients. They will also measure the effects of cortisol levels on immune function, which Dr. Lee Berk of Loma Linda University, whose work is reported in Chapter 10, has shown to be associated with decreased immunity.

Because of evidence that nerve growth factors contribute to the healthy functioning of nerve cells elsewhere in the body, Dr. Herbert Weiner of our UCLA Task Force hypothesizes that such factors play a role in the onset of Alzheimer's disease. Once degenerated, nerve cells do not regenerate. Dr. Weiner wonders whether it is possible that they degenerate because nerve-cell–maintaining growth factors are withdrawn, resulting in nerve-cell death and deterioration of the source of stimulation to blood vessels in the brain.

Dr. Weiner is measuring the content of nerve growth factor and nerve growth factor–inhibiting substances in the spinal fluid of patients with Alzheimer's disease, patients with other forms of dementia, and control subjects. The identification of mechanisms underlying Alzheimer's disease will contribute to our understanding of how the central nervous system affects behavior.

Dr. Jean Merrill, associate professor in residence in the Department of Neurology at the UCLA School of Medicine, found that Substance P, a neuropeptide secreted by the brain, can increase brain-cell production of immunoregulators interleukin-1 (IL-1) and tumor necrosis factor (TNF)—a tumor killer and immune cell activator. However, this stimulation occurs only in conjunction with an infectious triggering agent. Dr. Merrill concludes that the brain has built-in amplifying mechanisms for producing both IL-1 and TNF.

As a followup, Dr. Merrill has been determining the sources and role of IL-1 and TNF under three different cirumstances: (1) in the initial stages of an immune response, (2) in normal brain development, and (3) in a disease state that may influence brain cell function (e.g., multiple sclerosis and AIDS). This research will help clarify the means by which the nervous and immune systems interact in health and disease.

Dr. Merrill's hypotheses are supported by the work of Dr. Fergus Shanahan, assistant professor in the UCLA Department of Medicine, and Dr. Peter Anton, of the UCLA Department of Gastroenterology. Drs. Shanahan and Anton have been developing methods for identifying the immune cells in the gastrointestinal tract that have neuropeptide receptors, in order to ascertain whether neuropeptides can influence the onset or course of inflammatory bowel disease. In particular, they have been trying to identify immune cells possessing receptors for the neuropeptide, Substance P, suspected to be associated with inflammatory bowel disease. By documenting interactions between specific neuropeptides and immune cells, Drs. Shanahan and Anton hope to discover possible pathways by which inflammatory bowel disease is sometimes triggered or exacerbated.

A clearly accepted hypothesis is that genetic factors play a role in schizophrenia. Dr. Arthur Kling studied a family of eleven—including father, mother, and nine children. Four of the children had been diagnosed as being paranoid schizophrenic at the age of twenty or twenty-one. Detailed interviews and studies of all the family members suggested that schizophrenic disorders may be the result of a variety of causes, which may be precipitated by either internal or external factors (e.g., drug abuse, psychosocial stress, or viral infections at key stages of development—possibly contributing to alterations in the central nervous system).

Because schizophrenic siblings showed subnormal immune responses to the measles virus, Dr. Kling surmises that research examining immune system and brain neurotransmitter interactions may be useful in uncovering the origins of schizophrenia. This work corresponds to the research of Dr. George Solomon of the UCLA Task Force (who reported connections between schizophrenia and immune abnormalities in the 1960s), Dr. Karen Bulloch of the University of California at San Diego, and Dr. Roger Gorski of UCLA.

Immune cells have been found to possess a variety of receptors for hormones, suggesting that the neuroendocrine system plays a role in nervous and immune system intercommunication. Dr. C. Patrick Reynolds, assistant professor in residence of the Division of Cancer, Immunology, and Biology in the Department of Pediatrics at the UCLA School of Medicine, has been attempting to determine whether neurons and immune cells communicate directly or through some other means such as hormones. This research will also provide information on the regulatory purposes of nervous and immune system interactions. Finally, because neuroblastoma is a tumor of the sympathetic nervous system (which plays a major role in producing the heightened metabolic states associated with the "fight or flight" response to stressors), the discovery of such cellular interactions will add insight into how stress influences immunity.

Malnutrition is a complication of cancer and chemotherapy that can affect a patient's survival. Dr. David Heber, associate professor and chief of the Division of Clinical Nutrition in the Department of Medicine at the UCLA School of Medicine, has been investigating underlying reasons for rapid weight loss and debility ("cachexia") in cancer patients, which he has found to be related to metabolic abnormalities such as fluctuating insulin and blood sugar levels triggered by the cancer. Bodily resistance to insulin (a hormone that enables blood sugars produced from food ingestion to be stored and used as fuel for the body) causes the muscles to burn their own protein for energy. Treatment of cancer, therefore, requires more than a direct attack on cancer lesions. If cachexia could be reversed by dealing with the metabolic abnormalities, a stage might be set for more effective treatment of the cancer itself. Dr. Heber is testing the effectiveness of insulin therapy and other medications to regulate blood sugar levels in reversing cachexia in cancer patients. He emphasizes that the nutritional management of cancer patients should begin at the time of diagnosis.

The effect of relaxation techniques on the immune status of Chronic Fatigue Syndrome patients is being studied by Drs. Jacob Zighelboim of the UCLA Department of Microbiology and Immunology and Raphael Melmed, visiting professor at UCLA from The Hebrew University in Jerusalem. Preliminary analyses of the immune profiles of Chronic Fatigue Syndrome patients suggest that their killer-cell activity is altered. Drs. Zighelboim and Melmed have observed increases in NK-cell activity and in the number of cells of the NK-cell family. By contrast, the killing activity of activated macrophages was unaffected in this group of patients. Data on the effectiveness of relaxation techniques are currently being analyzed.

DOCTORS AND NUTRITION

Dr. Robert A. Good, former head of the Memorial Sloan-Kettering Cancer Center, and later at the University of South Florida, went to Africa to gather materials for his immunological studies. His attention was drawn to refugee children who had been undernourished since birth. They had been deprived of normal breast feeding; as a result of malnutrition, their mothers' milk had dried up. Dr. Good was struck by an evident connection between the persistent malnutrition of the children and the recurrent serious infections that produced a high mortality rate. The infections signaled deficient immune systems. On the same African trip, he observed other children who had enough to eat but whose diet consisted mainly of carbohydrates with very little protein. The lack of balanced food intake resulted in lowered immunity, manifested again by serious infections and a high death rate.

Dr. Good pursued these observations with several large-scale studies and was able to document the connection between faulty diet and impaired immunological function, which resulted in recurring susceptibility to a wide range of illnesses. When he investigated the cause of stunted growth, lethargy, and high frequency of infection in a certain strain of Holstein cattle, he discovered two things: first, a severe deficiency of the thymus gland, with consequent shortage of disease-fighting T cells; second, a genetic metabolic defect that resulted in an inability to absorb essential amounts of zinc from the gastrointestinal tract. Lacking zinc, the thymus gland became underdeveloped, with a corresponding underproduction of T cells. Once the cattle received zinc supplements in their feed, the various illnesses abated. Young cattle given zinc grew to full size.

The significance of these facts extends to human beings. People who have a shortage of zinc, whether as a result of faulty metabolism or because of inadequate dietary intake of zinc, are subject to a wide range of illnesses that might otherwise be prevented. Dr. Good observed that zinc-deficient children often had rashes around their orifices and on their extremities, and increased vulnerability to infections and various illnesses. Their immune systems either didn't produce enough natural killer cells or produced defective NK cells. Thymus hormone levels were low and the entire immune system was, in fact, grossly deficient. Yet when these children were given sufficient amounts of zinc, the immune system began to function properly.

Dr. Good extended his studies to cancer patients. Here, too, he found a frequent coincidence of zinc deficiency. Obviously, Dr. Good didn't prescribe zinc as a "cure" for cancer. Nonetheless, the impairment of immune function resulting from insufficent zinc should not be ignored in any comprehensive strategy of treatment. Dr. Good is unequivocal in his statement that zinc is a "critical element in immunological function." He points out that, apart from cancer, zinc deficiency is often found in illnesses involving inflammatory disease, alcoholism, exogenous drugs, or toxicity. Older people, too, he finds, tend to suffer from zinc deficiency.

It would be absurd, of course, to implicate inadequate levels of zinc as the sole or even the main cause of any serious disease. It is not wrong, however, to talk about the role of nutritional deficiencies in contributing to such disease, including cancer.

And nutritional problems involve not just shortages but too much food. Americans, on the whole, suffer more from overeating than from shortages. The amount of fat in the average American diet is 40 percent, or about twice as much as is required for good health. The main effect is apt to be shortened life expectancy marked by increased susceptibility to a wide range of serious illnesses, all the way from heart disease and arthritis to kidney malfunction.

Few things are more important for good health than a good circulatory system that carries nutrients to every part of the body. But the wrong foods can block arteries and blood vessels, narrowing the blood vessels, reducing the amount of oxygen available to the heart, and forcing it to pump blood through constricted passageways.

Generally, the amount of cholesterol deposited in the arteries is used as an index of faulty diet and therefore of risk from heart attack. An equally significant figure applies to LDL and HDL—low-density and high-density lipoproteins. A high level of LDL is considered to be dangerous, whereas a high HDL is considered a good sign.

Until the 1980s, most physicians associated abnormally high levels of cholesterol solely with diet. More recent research, however, has established a connection between emotional stress or physical fatigue and abnormally high cholesterol. As mentioned elsewhere in this book, high cholesterol levels have been found in accountants approaching IRS deadlines, or in medical students approaching examinations. High cholesterol counts, whether from food or emotional stress, can represent an invitation to heart attacks, but, as Dr. Good and his associates have found, they can be a portal for a wide range of other illnesses as well, since a heavy burden of fat can do damage to the immune system. Obesity and overeating and a healthy immune function, he says, do not go together. There is no doubt in Dr. Good's mind of the link between faulty nutrition and an impaired immune system.

Dr. Good's observations were confirmed and reinforced by the work of Dr. Roy Walford of UCLA, whose detailed findings make clear that one of the quickest ways to a short life is through overeating. Dr. Walford's experiments have also demonstrated that the simple act of cutting down on food intakes in animals can produce a significant increase in life expectancy. Even cancer has been linked to overeating. Dr. Good contends that seriously overweight women have ten times more chance of developing malignancy than women of normal weight.

Children who are neither undernourished nor overfed tend to be stronger and brighter than their counterparts. Healthy people of old age whose immune systems are functioning at a high level are generally found to be those who eat wisely, maintaining a good balance between carbohydrates and protein.

In recent years, as people have become increasingly conscious of the connection between nutrition and health, physicians have been accused of giving too much emphasis to medications and not enough to foods. A related complaint is that not until nutrition is given status in medical schools will doctors place a healthy diet on the same essential level as drugs.

Physicians will respond that many people allow themselves to be victimized by health-food hucksters and faddists. These physicians argue that hundreds of millions of dollars are squandered on vitamins and other food supplements. Many of these same physicians contend that the average supermarket shopping basket contains everything required for a balanced diet. Nutritionists fight back by saying that processed foods are so devitalized and the pollutants in the environment so profuse that consumers are forced to augment their usual diets. It is said that some physicians don't take these environmental factors into account.

Dr. C. Everett Koop, then the U.S. surgeon general, in a special 1988 report on nutrition and disease, disagreed with those physicians who contend that the average supermarket shopping basket provides for general nutritional needs. Dr. Koop supported a host of studies showing that most people are eating too much dietary fat, thus increasing their risk of heart disease, diabetes, and strokes. His report found that improper diet was associated with five of the ten leading causes of death.

Almost eighty years ago, when questions were raised about the adequacy of medical education, an impartial study was undertaken under the auspices of the Carnegie Foundation. The principal investigator was Abraham Flexner, whose report revealed drastic shortcomings in medical education and which served as the basis for far-reaching reforms.

It seemed to me that the best way of resolving the argument about the physician's knowledge of nutrition was to investigate, in a Flexner-type study, the curricula of the nation's medical schools. As a member of the board of trustees of the Ruth Mott Fund of Flint, Michigan, and as chairman of its Health Committee, I proposed in 1984 that the Fund initiate such a study, and seek the cooperation of the Food and Nutrition Board of the National Research Council.

The Mott board approved the proposal and appropriated $200,000 for the study. James Kettler, the executive director of the Ruth Mott Fund, wisely suggested that we seek funding partners because of the possibility that a far-reaching study might cost more than the amount allotted to the board.

The board encouraged Jim Kettler to explore all the possibilities. Within a few weeks, Jim had everything in place. The William H. Donner Foundation agreed to serve as cosponsor. The Food and Nutrition Board, recognizing the value of such a study, agreed to mount the actual study, and appointed a Committee on Nutrition in Medical Education. Committee members were: Myron Winick, chairman, Institute of Human Nutrition, College of Physicians and Surgeons, Columbia University; Stanley M. Aronson, Division of Biology and Medicine, Brown University; Richard Behrman, Department of Pediatrics, School of Medicine, Case Western Reserve University; Lucille S. Hurley, Department of Nutrition, University of California at Davis; Douglas S. Kerr, Department of Pediatrics and Biochemistry, Case Western Reserve University; Alexander Leaf, Department of Preventive Medicine and Clinical Epidemiology, Harvard Medical School; J. Michael McGinnis, Office of Disease Prevention and Health Promotion, U.S. Department of Health and Human Services; and Jacqueline Ann Reynolds, Physiology Department, Duke University.

The committee proceeded to examine the curricula of the nation's medical schools, scrutinizing not just nutrition teaching per se but related subjects in which nutrition was included. The committee also looked at the level of knowledge about nutrition possessed by graduating medical students.

I can think of few more important public documents relating to health than the report produced by this study, issued in 1985.

Herewith, pertinent excerpts from the "Conclusions and Recommendations" of the report:

> The committee concluded that nutrition education programs in U.S. medical schools are largely inadequate to meet the present and future demands of the medical profession. This perception, reflecting results of prior surveys and conferences, was confirmed by the committee's independent survey and related investigations as outlined above.
>
> The committee recommends that medical schools and their accreditation bodies, federal agencies, private foundations, and the scientific community make a concerted effort to upgrade the standards as detailed below. The committee recognizes the extraordinary demands placed on the medical education system of today. Nevertheless, it believes these changes could be achieved with minimal disruption of existing curricular and administrative structures, although in most cases this upgrading may require philosophical adjustment.
>
> The committee recommends to medical schools that the basic principles of nutrition be introduced simultaneously with other preclinical sciences as an independent course, and that the precepts of nutrition be reinforced later during clinical training to demonstrate their application to patient care. This recommendation stems from the recognition that the importance of nutrition is not sufficiently recognized by the faculty and that its impact is significantly diminished when it is not taught as a discrete entity. The present survey demonstrated that most schools teach some nutrition in one form or another; however, only two-thirds of them teach it in the first academic year and approximately 20 percent teach nutrition as a separate, required course. In medical schools, elective courses are distributed throughout the four years of the basic science curriculum and range in duration from less than four weeks to more than ten weeks. Although many schools offer nutrition clerkships or electives, the results of both the

committee's and the AAMC's [Association of American Medical Colleges] surveys indicated that only a small segment of the student body takes advantage of these options. In contrast, required courses serve as a focal point for a discipline and significantly increase the probability that the student body has a uniform base of knowledge.

The committee proposes that the following topics in nutrition become part of the basic curriculum of medical schools and, furthermore, that they be integrated into clinical clerkships: energy balance, role of specific nutrients and dietary components, nutrition in the life cycle, nutritional assessment, protein-energy malnutrition, the role of nutrition in disease prevention and treatment, and risks from poor dietary practices stemming from individual, social, and cultural diversity. To cover these core concepts adequately, a minimum of 25 to 30 classroom hours should be allocated to them during the preclinical years. At present, only 21 hours, on average, are given to these subjects. According to the committee's survey, there is great variation in the number of nutrition hours taught. For example, approximately 60 percent of the schools surveyed provide less than 20 hours in nutrition instruction and 20 percent teach less than 10 hours. Only 30 percent teach 30 or more hours. The number and distribution of hours devoted to nutrition in clinical settings could not be determined with any degree of confidence.

The committee's survey also demonstrated considerable variation in the scope of topics included in nutrition courses. Some subjects, such as energy balance and obesity, are covered in almost all schools, whereas others, such as the role of nutrition in health promotion and disease prevention, receive attention in only a few schools.

The committee recommends that persons with strong backgrounds in nutrition science, research, and applications to clinical medicine be assigned to lead the development of nutrition programs in medical schools. Physician-nutritionists, well versed in the clinical application of basic research, would be the ideal candidates. Currently, faculty leadership for nutrition programs is shared by M.D.'s and Ph.D.'s. In the nine schools identified through the survey as having well-established programs, M.D.'s play a strong, central role in teaching nutrition and in demonstrating its application to clinical medicine. The committee encourages medical schools to involve M.D.'s as well as Ph.D.'s in the instruction of nutrition.

There are variations in the administrative structure of U.S. medical schools and, as a consequence, differences in faculty responsibility. Because authority for nutrition education is often not centralized, the success of a program often depends heavily on individual initiative. In approximately 80 percent of the schools that teach nutrition, responsibility is shared by scholars engaged in basic sciences such as biochemistry, physiology, and pharmacology, or in clinical disciplines such as pediatrics, medicine, surgery, and gastroenterology. Faculty coordinators interviewed by the committee on average devoted 40 percent of their time to nutrition research. The committee determined, however, that although faculty training in nutrition science appears to be minimal in schools that place little emphasis on nutrition, renowned nutrition scientists in several medical schools do not seem to be engaged in teaching nutrition at their institutions [either]. A strong research program in nutrition enhances the credibility of the faculty and provides financial security, but nutrition research per se in medical schools does not guarantee that nutrition will be taught.

To ensure permanence of the nutrition programs, the committee recommends that responsibility for the programs be vested in a separate department of nutrition or in a distinct clinical division of the medical school. Moreover, it recommends that each institution allocate funds for the support of at least one faculty position in nutrition. At present, faculty positions specifically designated for nutrition are rare. The financial burden attendant on meeting these goals may be partially offset by income generated from nutrition-related clinical support services within the hospital, but in the immediate future will have to be derived predominantly from research support or other sources.

The committee supports the concept of diverse approaches in medical education and recognizes that each school must devise its own curriculum design, implementation strategy, and organizational structure. In its judgment, however, lack of organizational structure and administrative and financial support are the prime hindrances to the maintenance of nutrition programs in medical schools.

The committee proposes that the National Board of Medical Examiners consider appointing advisers to review the distribution and quality of nutrition-related questions on board examinations and to establish a mechanism for communicating such findings and recommendations to board section chairmen. Such advisers could also identify areas of clinical nutrition that deserve coverage in the examination or provide new questions for consideration by the board committees. Of the approximately 6,000 examination questions reviewed by the committee, 3 percent to 4 percent had some relation to nutrition and the distribution of the questions on nutrition among the medical specialties was noticeably even. There were no questions on several topics deserving emphasis, e.g., osteoporosis, nutritional requirements of the elderly, total parenteral and enteral feeding techniques, and the relationship between nutrition and cancer.

The committee recommends that the federal government and private foundations provide additional financial support for the development of teaching aids and the training of a cohort of clinical scientists with competence in nutrition. Nutrition coordinators and faculty reported that the resources for teaching nutrition in medical schools are insufficient and that nutrition textbooks and ancillary aids, although plentiful, are unsuited to their needs. Both these deficits place additional demands on medical schools and are a deterrent to the development of nutrition programs. Although the committee encourages institutional sharing of faculty and resources as an interim measure, the long-term survival of nutrition programs is dependent on increases in funds from federal and private sources.

To evaluate existing programs more accurately and to assist in planning for the future, the committee recommends that a mechanism be established to monitor periodically changes in the status of nutrition education in medical schools. One device would be to include more exploratory questions on nutrition in the annual survey of medical school curricula conducted by the AAMC. These questions should be directed at persons with primary responsibility for the program. The committee also encourages each medical school to monitor its own program to ensure that it remains abreast of advances in nutrition. Finally, the committee recommends that in approximately five years, an authoritative body such as the Food and Nutrition Board of the National Research Council reexamine the status of nutrition in U.S. medical schools. The absence of a reliable surveillance mechanism thus far has severely hampered the ability to define the dimensions of the problem and to characterize progress.

All the elements outlined above—placement of nutrition in the curriculum, scope and duration of courses, financial and administrative support for faculty and research, attention by accreditation bodies that influence medical education, and mechanisms for monitoring progress—are essential to ensure that nutrition programs are initiated and that existing ones are rejuvenated and sustained. The committee recognizes the difficulties that attend any curricular change in undergraduate medical education. Nonetheless, it believes that most medical schools could implement the above recommendations now without major reallocation of funds or displacement of other disciplines.

None of my involvements over the years in medical matters has been more personally gratifying than to have had a part in initiating this study by the top scientific body in the country, documenting the need for medical schools to raise the level of nutrition education. I can think of few changes in the curriculum of medical schools that might contribute more to the health of the American people. The report serves as a virtual mandate for physicians to question patients carefully on their food habits as possible clues to illness, as well as to regard nutrition, and not just medication, as a vital aspect of treatment.

TO THE GRADUATES: A PHYSICIAN'S CREDO

It is common for graduating medical students to select their own commencement speaker. I was especially pleased, therefore, when Dean Mellinkoff informed me, after I had been at the school for less than a year, that he was conveying the invitation of the senior class to speak at its graduation.

The main and understandable desire of graduating students is to feel the diplomas in their hands and march out as soon as possible. This desire tends to increase in direct proportion to the temperature, usually a factor in late May or early June. A commencement speaker, therefore, carries the natural liability of standing between the graduates and their immediate aspirations. The possibility that a commencement speaker will let loose new truths on seasoned ears is not very high.

The following text embodies material drawn from commencement talks at the medical schools of UCLA, Harvard University, George Washington University, and Baylor University.

> There are qualities beyond pure medical competence that patients need and look for in doctors. They want reassurance. They want to be looked after and not just looked over. They want to be listened to. They want to feel that it makes a difference to the physician, a very big difference, whether they live or die. They want to

feel that they are in the doctor's thoughts. The physician holds the lifeline. The physician's words and not just his prescriptions are attached to that lifeline.

This aspect of medicine has not changed in thousands of years. Not all the king's horses and all the king's men—not all the tomography and the thallium scanners and two-D echocardiograms and medicinal mood modifiers—can preempt the physician's primary role as the keeper of the keys to the body's own healing system.

I pray that you will never allow your knowledge to get in the way of your relationship with your patients. I pray that all the technological marvels at your command will not prevent you from practicing medicine out of a little black bag. I pray that when you go into a patient's room you will recognize that the main distance is not from the door to the bed but from the patient's eyes to your own—and that the shortest distance between those two points is a horizontal straight line—the kind of straight line that works best when the physician bends low to the patient's loneliness and fear and pain and the overwhelming sense of mortality that comes flooding up out of the unknown, and when the physician's hand on the patient's shoulder or arm is a shelter against the darkness.

I pray that, even as you attach the highest value to your science, you will never forget that it works best when combined with your art, and, indeed, that your art is what is most enduring in your profession. For, ultimately, it is the physician's respect for the human soul that determines the worth of his science.

In an age in which honest labeling becomes the starting point for any transaction, it is probably a sound idea for the physician to present his or her credo and define the governing principles he intends to bring to the relationship with the patient. Such a credo can provide an accurate and sensible basis for the patient's anticipations.

Presumptuously, I should like to propose one such credo. I do so only for illustrative purposes. I suspect that there may be as many credos as there are physicians.

TO MY PATIENTS:

It may be helpful, at the outset of our relationship, to tell you how I propose to get you well and to keep you that way.

The first thing to be emphasized is that you own a body that has been beautifully honed by at least a million years of experience in coping with all sorts of disorders. This built-in coping capacity and cellular wisdom goes by the name of the immunological system. I have no hesitation or loss of pride in admitting that I know far less about you than your own body does. My job is to put that natural knowledge fully to work when anything goes wrong. I am not the healer. You are. My job is to help activate and accelerate the healing mechanism.

The science of medicine and the art of medicine come together when three things are achieved. The first is the accuracy of the diagnosis. The second is the proportionate nature of the treatment. The third is the full mobilization and release of your own healing resources in which your robust expectations of full recovery can play such an important part. If I can orchestrate these three elements, I will be doing what I am supposed to do.

Let me say here that my greatest difficulty in dealing with you will not be when you are really sick but when you are really well and think you are really sick. I have in mind those times when you will come to me with pains, severe pains in many cases, for which I can find no organic cause. So there you will be, sitting in my examining room with a squeezing pain in your gut or with a headache, and there I will be, standing over you and having the nerve to tell you your physical examinations show nothing wrong. At that point, you will probably feel like getting up, walking out, and going across the street to one of my colleagues in the hope that he can really give you something to worry about.

I want to anticipate that situation now by saying that I will not be accusing you of imagining your symptoms or of being a hypochondriac. I won't doubt that those pains are real. I have had such pains myself. And the fact that the pains are not caused by a dreaded disease, which understandably is a question at the back of your mind, does not mean that they don't exist. What it probably means is that your body has been trying to tell you something and you haven't been listening.

It may be trying to tell you that your circuits are badly overloaded and that you have to cut down on high-voltage tensions—whether they originate in your office or your home or anywhere else.

It may be telling you that you are no more adept at handling exasperations and frustrations than anyone else, or it may be telling you you are not as much of a speed king as you like to think. There is a great deal of pressure these days to be a fast jumper, whether with respect to businesses, beds, beliefs, or buddies.

Maybe it is no more complicated than the fact that you are spending too much time on the freeways. We live in an age when total mobility provides us not with total access but with total congestion.

Maybe your body is protesting at having to carry more caloried cargo than it can use or than its frame was intended to support—cargo that has bumped its way down a well-worn and overtrafficked alimentary canal. Your desire for food and your need for food may be wildly out of sync.

Maybe your body is telling you that you are mistaken if you want to prove you can be the first human being in the history of the race who can live without an adequate oxygen supply. You cannot deliberately and persistently saturate your lungs with smoke and still expect them to behave as though you have been taking deep-breathing exercises in the Swiss Alps.

These may be only a few of the reasons why your body is using pain to get your attention. While I may try to reassure you that those pains are not connected to a disease, I will not minimize the need to locate the underlying trouble. We will have to look into every corner of your life to find out why your warning signals are flashing.

Most people think they are going to live forever—until they develop a cold, when they think they are going to die within the hour. They tend to regard pain as the ultimate enemy and not as a respectful suggestion that they sit still and think about the way they are insulting their own bodies. You will only add to these insults if you shovel down aspirins or other analgesics. They may temporarily block off the sensations of pain but they may also be intensifying the underlying condition. If you really need analgesics, I want to be the one to prescribe them.

I emphasize the subject of pain because your own attitude toward pain will have a great deal to do with your health. People who worry to the point of panic because of sudden, sharp pain invite serious danger. Panic is a fierce multiplier of disease. If you should ever find yourself in a situation where you are suddenly assailed by more intense pain than you have ever experienced before, I beg you to make a supreme effort not to let that pain scare the life out of you. Just have confidence in the fact that competent help is on the way and that the combination of medical science and the wisdom of your own body, to use Walter Cannon's wonderful phrase, is a powerful team and knows what to do.

Meanwhile, I will do my best to protect you against panic-causing situations on my end. In the remote event, for example, that it should ever be necessary for you to be sent to an intensive-care unit, I want to be on hand when you arrive, so that you will not be plunged into a strange and intimidating environment of electronic calibrating devices being hooked into your vital impulses. I want to be there so you will understand what is happening and to take some of the terror and depersonalization out of your encounter with calamity-measuring instrumentation.

The same is true of surgery. The world of the operating room, when viewed from the table, takes on a terrifying perspective. A patient looks up at big bright lights, uniforms, and masks. If this is the last thing he sees before the black rubber mouthpiece goes on, then this is not a satisfying or sustaining image for the subconscious to carry through surgery. But when the patient knows that his own doctor is near him, and can feel a compassionate hand on his shoulder, his apprehensions are eased and his confidence nourished.

In short, at critical times, I intend to be at your side. If ever I should be too busy to do this, I will know that the time has come for me to haul in my shingle, because there is no more important mandate to me as a physician than to be with my patient when he or she really needs me.

In any illness we both have work to do. So far, I have been describing my job. Now let me describe yours. The more serious the illness, the more important it is for you to fight back. You've got to mobilize all your resources—spiritual, emotional, intellectual, physical. Your heaviest artillery will be your will to live. Keep that big gun going.

And, incongruous though it may sound, if you've got any capacity for joy, this will be the time to put it to the fullest use. It makes no sense to believe that emotions have an effect on the body's chemistry only when they're on the downside. Every emotion, negative or positive, makes its registrations on the endocrine system.

Illness is not a laughing matter. Maybe it ought to be. Laughter is a form of internal jogging. It moves your internal organs around. It enhances respiration. It is an igniter of great expectations. Your body will experience a powerful gravitational pull in the direction of those expectations. Your hopes are my secret weapon. They are the hidden ingredient in any prescription I might write. So I will do everything I can to generate and encourage your confidence in yourself and in the certainty of recovery.

This is my notion of what a partnership between physician and patient is all about.

Sincerely yours,
Your Doctor

A GUIDE TO TECHNICAL TERMS

Acquired immune deficiency syndrome (AIDS): A disease caused by the human immunodeficiency virus (HIV) and characterized by the lack of helper T cells and the susceptibility to opportunistic infections and certain tumors.

Antibodies: Proteins produced by plasma cells evolved from B lymphocytes in response to antigen stimulation. These may combine with and neutralize microorganisms, or facilitate their destruction by other cells. Found in blood, body fluids, and tissues. All antibodies are immunoglobulins.

Antibody titers: See *titers*.

Antigen: Toxin, foreign protein, bacteria, or tissue cells capable of inducing a protective response by the immune system.

Autoantibody: An antibody that attacks the body's own tissue.

Autoimmune disease: Conditions related to the immune system acting against the body's own tissues.

Autonomic nervous system: Part of the nervous system that regulates cardiac muscle, smooth muscle (found in the walls of hollow organs), and glandular tissues.

Bacterium: An organism discernible by microscope that is composed of a single cell. Most bacteria are complete cells capable of replicating themselves. Often disease-causing. Plural: *bacteria*.

B cells: Immune cells derived from bone marrow that are precursors of plasma cells that produce antibody to protect against infection. B-cell overactivity can lead to exaggerated physiological reactions, as in asthma, and occasionally B cells can produce antibodies that attack the body's own tissues (as in systemic lupus erythematosus, rheumatoid arthritis). Also called *B lymphocytes.*

Biology: The study of organisms.

Biological: Relating to the understanding of life and living processes.

Blastogenesis (a.k.a. lymphocyte activation, proliferation, responsiveness, stimulability): Immune cell growth and division following exposure to *antigen* or *mitogen.*

B lymphocytes: See *B cells.*

Bone marrow: The soft tissue filling the cavities of bones. Site of immune cell production.

Chronic Fatigue Syndrome: Chronic symptoms of fatigability, depression, loss of appetite, insomnia, and inability to concentrate, without an identifiable physical cause.

Clinical: Direct work with patients or the condition of patients.

Conditioned reflexes: The physiological response provoked by a conditioned stimulus.

Conditioned stimulus: Any sensory experience or object that will trigger an unrelated physiological response after having been associated with something that actually elicits that physiological response.

Corticosteriods: A class of steroid hormones, secreted by the adrenal gland and serving an anti-inflammatory and energy-conserving function (e.g., cortisol and corticosterone).

Cortisol: See *corticosteriods.*

Cytotoxic T cells: T cells that, upon activation by a specific antigen, target and attack the cells bearing that type of antigen. Also called Killer T cells.

Endocrine system: The glandular system that secretes hormones into the blood to influence metabolism and other body processes (e.g., histamine affects small-vessel blood flow to tissues, air-passage regulation, and gastric secretion). Susceptible to changes in emotion.

Endorphins and enkephalins: Brain secretions that have analgesic or pain-inhibiting effects.

Epinephrine (a.k.a. adrenaline): A hormone in the chemical family of catecholamines and powerful stimulator of the sympathetic nervous system (for example, increasing blood pressure and heart rate, and heightening metabolic activity). Also a neurotransmitter. See *sympathetic nervous system*.

Epstein-Barr virus (EBV): Virus associated with the development of infectious mononucleosis and other fatigue-related syndromes.

Fight-or-flight response: A term developed by Dr. Walter Cannon to describe a heightened metabolic state associated with stress, when the body is physiologically geared to fight or flee from a threatening situation.

Helper T cells: T cells that prompt B cells to form antibodies and that secrete interleukin-2 to facilitate cytotoxic-T-cell proliferation and activity.

Herpes simplex virus (HSV): Virus associated with the development of herpes simplex, an acute viral disease. HSV-1 infection is typified by clusters of vesicles on the skin, often on the borders of the lips and nostrils (in the form of cold sores). HSV-2 infection is largely transmitted sexually and usually involves lesions on genital and surrounding areas.

HIV (human immunodeficiency virus): Human virus that is the cause of AIDS.

Hormone: A substance produced by the endocrine system that regulates specific cellular or organ activity.

Hypertension: High blood pressure.

Hypothalamus: The portion of the brain that regulates and coordinates the endocrine system, the immune system, moods, motivational states, and behavior. Controls the pituitary gland, a major gland of the endocrine system, which is a part of the hypothalamus. A main component of the limbic system.

Iatrogenic: Said of any adverse condition brought about as a result of any aspect of medical treatment, including hospital care.

Immune system: The organ system (predominantly consisting of the spleen, bone marrow, lymph nodes, and lymphoid cells circulating throughout the body) that protects the body from harmful foreign substances such as bacteria, viruses, and malignant cells.

Immunocompetence: Effectiveness of the immune system; immune function.

Immunoglobulins: Antibody activity is the property of immunoglobulin molecules. There are five major classes of immunoglobulins: IgA, IgD, IgE, IgG, and IgM. Each immunoglobulin targets specific antigens. The whole immunoglobulin population in each person reacts with an enormous variety of antigens.

Immunoglobulin A (IgA): One of the five major classes of immunoglobulins, or antibodies, which protects outer surfaces of the body. Found in mucous secretions of the intestinal and respiratory tract, and in colostrum and milk.

Interferons: Lymphokines that protect noninfected cells from viral infection and that augment natural-killer-cell, cytotoxic-T-cell, and macrophage activity.

Interleukins: Lymphokines that regulate immune cell growth, maturation, and function.

Lesion: An injury or tissue abnormality such as a wound, ulcer, or malignant growth. Also, an individual part (e.g., a nodule) of a multifocal disease.

Leukocytes: White blood cells, or types of major effector cells in the immune system, consisting of neutrophils, eosinophils, basophils, monocytes, and lymphocytes. The other major types of immune cells are plasma cells and macrophages.

Limbic system: A section of the brain that is associated with our sense of smell, involuntary functions, emotions, and behavior. The hypothalamus is considered to be a main component of the limbic system.

Lupus: See *systemic lupus erythematosus*.

Lymph: A second circulatory system containing lymphocytes that flows through all tissues and empties into the blood circulation.

Lymph nodes: Bean-sized organs located throughout the body in the route of lymph flow in which many immune cells are formed and stored.

Lymphatic organs: Organs (e.g., thymus) that contain lymphocytes. Also called *lymphoid tissues*.

Lymphocytes: Cells of the immune system. Genetically programmed to recognize specific antigens that bind to the lymphocyte and stimulate the development of more lymphocytes reacting with those antigens. T and B cells comprise the two main lymphocyte subpopulations.

Lymphocyte proliferation: See *blastogenesis*.

Lymphocyte responsiveness: See *blastogenesis*.

Lymphocyte stimulability: See *blastogenesis*.

Lymphokines: A general term including all nonantibody chemical messengers (e.g., interleukins, interferons) secreted by lymphocytes that play a major role in regulating immune reactions. Also produced by other cells of the body.

Macrophages: A class of immune cells capable of engulfing and digesting foreign or toxic matter within bodily tissues, and performing accessory functions in immune-cell activation.

Malignant melanoma: See *melanoma*.

Melanoma: A form of skin cancer that has a marked tendency to metastasis. Also called *malignant melanoma*.

Metastasis: Transfer of cancer from one organ or part of the body to another not directly connected with it.

Microorganism: A microscopic or submicroscopic organism (e.g. bacterium or virus).

Mitogen: A substance foreign to the body that induces lymphocytes to proliferate. Different mitogens stimulate different subpopulations of lymphocytes.

Molecule: A particle of chemical substance.

Monocytes: Circulating cells that evolve into macrophages once they enter bodily tissues. Possess the ability to migrate to sites of inflammation and transform into macrophages.

Naloxone: A synthetically produced endorphin-blocking agent.

Natural killer (NK) cells: A class of cells that possess the spontaneous ability to recognize and kill tumor- and virus-infected cells without previous exposure to the antigen. Critical to initial defense against infections, especially viral.

Nervous system: The system involving the brain and spinal cord that, along with the endocrine system, coordinates adjustments and reactions of the mind and body, affecting voluntary and involuntary muscles, nerves, and glands throughout the body.

Neuroendocrine: Relating to the shared influence of and interactions between the nervous and endocrine systems, including the production of hormones.

Neuroimmunomodulation: The study of the mechanisms within the nervous system involved in the regulation of immunity.

Neuron: Nerve cell.

Neuropeptides: Originally defined as hormonelike substances, produced by nerve cells. Also used as a general term describing classical neuropeptides (such as endorphins) and substances studied in other contexts such as hormones, lymphokines, and growth factors, that carry messages to the brain and throughout the body in order to control a wide variety of bodily functions. Neuropeptides pass messages by attaching themselves to receptors on cell walls. Produced by ncurons, lymphocytes, and other cells of the body.

Neurotransmitters: Substances produced and released by nerve cells to stimulate activity in susceptible cells (e.g., nerve cells, organ cells).

Norepinephrine (a.k.a. noradrenaline): A hormone in the chemical family of catecholamines and a major neurotransmitter of the sympathetic nervous system that is closely related to epinephrine in its stimulating influence on cardiovascular and other related functions. See *epinephrine* and *sympathetic nervous system*.

Pathogen: Any disease-producing agent or microorganism.

Placebo: An innocuous substance or procedure having no particular physiological impact. Usually administered for purposes of satisfying a patient's psychological need for medication, comparing the actual effectiveness of a drug or procedure, or for determining the power of expectations in producing physiological change.

Psychoneuroimmunology: The study of how psychological and emotional states influence disease resistance via interactions with the nervous, endocrine, and immune systems. Also called *psychobiology, psychoimmunology*.

Psychosocial: Psychological and social.

Psychosomatic symptoms: Bodily symptoms related to psychic, emotional, or mental states.

Psychotherapy: Psychological techniques used to alleviate mental or emotional distress.

Receptors: Sites in the walls of cells that are geared to respond to particular agents (such as a neuropeptide or an invading virus).

Salivary immunoglobulin A (sIgA): Immunoglobulin A that is found in saliva. See *Immunoglobulin A.*

Somatization: The conversion of mental states into bodily symptoms.

Spleen: Large, glandlike immune organ in the abdominal cavity that is a major site for antibody production. Processes red blood cells and also serves as a blood reservoir.

Suppressor T cells: T cells that specifically inhibit antibody production by B cells as well as other cytotoxic-T-cell activities.

Sympathetic nervous system: A component of the autonomic nervous system involved in producing the physiological manifestations of the "fight or flight" response. Both epinephrine and norepinephrine are major substances involved in sympathetic nervous system activity. See *fight-or-flight response, epinephrine,* and *norepinephrine.*

Synapses: Nerve cell junctures where biochemical substances are passed.

Systemic lupus erythematosus: An acute or chronic inflammatory disease involving many organs that is associated with the production of autoantibodies against the body's own tissues.

T cells: Immune cells processed by the thymus that protect against viral, fungal, and intracellular bacterial infection; foreign (e.g., transplanted) and malignant cells. Consist of a variety of subtypes, including cytotoxic T, helper T, and suppressor T cells. Also called *T lymphocytes.*

Thymus: A glandlike central lymphoid organ located in the chest that is responsible for the development of T lymphocytes.

Titers: The quantity of a substance required to react with a given amount of another substance. Decreases in antibody titers to inactive viruses (e.g., to Epstein-Barr or herpes simplex viruses) are thought to reflect better immune control of the virus; whereas, viral activation results in increases of antibody titers for the virus.

T lymphocytes: See *T cells.*

Virus: An infectious organism that can only reproduce within living cells.

NOTES

1. Birth of an Obsession

Friedrich Bidder and Carl Schmidt published their observations regarding gastric changes accompanying emotional conditions in F. H. Bidder and C. Schmidt, *Die Verdauungssafte und der Stoffwechsel* (Mittauund Leipzig: G. A. Reybar, 1852).

Walter Cannon, *Bodily Changes in Pain, Hunger, Fear and Rage* (New York: W. W. Norton, 1963).

Walter Cannon, *The Wisdom of the Body* (New York: W. W. Norton, 1939).

Walter Cannon, *The Way of an Investigator* (New York: W. W. Norton, 1945).

Fritz Mohr's statement about the unity of psyche and soma can be found in Cannon, *Bodily Changes*.

Lewis Thomas, *The Medusa and the Snail* (New York: Viking Press, 1979).

2. Encounters with Patients

Walter Cannon, *Bodily Changes in Pain, Hunger, Fear and Rage* (New York: W. W. Norton, 1963).

Hans Selye's studies on the physiological effects of stress are described in *The Stress of Life* (New York: McGraw-Hill, 1956).

Norman Cousins, *Anatomy of an Illness* (New York: W. W. Norton, 1979).

3. Learning from Medical Students

The neighborhood survey that investigated the reasons for changing physicians was reported in an article by Norman Cousins, "How patients appraise physicians," *New England Journal of Medicine* 313 (1985): 1422–1425.

The complex role of the immune system is outlined by Dr. Gustav J. V. Nossal in "Current concepts: Immunology," *New England Journal of Medicine* 316 (1987): 1320–1325.

The scientific evidence providing the rationale for Dr. J. Edwin Blalock's conclusions about the sensory role of the immune system and the nervous-endocrine-immune system network is detailed in the following publications: J. E. Blalock, "The Immune system as a sensory organ," *Journal of Immunology* 132 (1984): 1067–1070. J. E. Blalock, "A molecular basis for bidirectional communication between the immune and neuroendocrine systems," *Physiological Reviews* 69 (1989): 1–32.

Franz Ingelfinger's article entitled "Arrogance," on the self-limiting nature of most human illnesses, appeared in the *New England Journal of Medicine* 303 (1980): 1506–1511.

Reports of accountants and students experiencing elevated cholesterol levels in response to emotional stress appeared in the following articles: M. Friedman, R. H. Rosenman, and V. Carroll, "Changes in the serum cholesterol and blood clotting time in men subjected to cyclic variation of occupational stress," *Circulation* 17 (1958): 852–861. S. M. Grundy and A. C. Griffin, "Effects of periodic mental stress on serum cholesterol levels," *Circulation* 19 (1959): 496–498. C. B. Thomas and E. A. Murphy, "Further studies on cholesterol levels in the Johns Hopkins medical students: The effect of stress at examinations," *Journal of Chronic Diseases* 8 (1958): 661–668. S. Wolf, W. R. McCabe, J. Yamamoto, C. A. Adsett, and W. Schottstaedt, "Changes in serum lipids in relation to emotional stress during rigid control of diet and exercise," *Circulation* 26 (1962): 379–387.

The many studies by Drs. Ronald Glaser and Janice Kiecolt-Glaser that demonstrate the link between negative emotional states and hampered immunity are: R. Glaser., J. K. Kiecolt-Glaser, C. E. Speicher, and J. E. Holliday, "Stress, loneliness, and changes in herpes virus latency," *Journal of Behavioral Medicine* 8 (1985): 249–260. R. Glaser, J. K. Kiecolt-Glaser, J. C. Stout, K. L. Tarr, C. E. Speicher, and J. E. Holliday, "Stress-related impairments in cellular immunity," *Psychiatry Research* 16 (1985): 233–239. R. Glaser, J. Rice, C. E. Speicher, J. C. Stout, and J. K. Keicolt-Glaser, "Stress depresses interferon production concomitant with a decrease in natural killer cell activity," *Behavioral Neuroscience* 100 (1986): 675–678. R. Glaser, J. Rice, J. Sheridan, R. Fertel, J. Stout, C. Speicher, D. Pinsky, M. Kotur, A. Post, M. Beck, and J. K. Keicolt-Glaser, "Stress-related immune suppression: Health implications," *Brain, Behavior and Immunity* 1 (1987): 7–20. J. K. Kiecolt-Glaser, W. Garner, C. Speicher, G. M. Penn, J. E. Holliday, and R. Glaser, "Psychosocial modifiers of immunocompetence in medical students," *Psychosomatic Medicine* 46 (1984): 7–14. J. K. Kiecolt-Glaser, R. Glaser, E. C. Strain, J. Stout, K. Tarr, J. Holliday, and C. Speicher, "Modulation of cellular immunity in medical students," *Journal of Behavioral Medicine* 9 (1986): 5–21. J. K. Kiecolt-Glaser, R. Glaser, E. C. Shuttleworth, C. S. Dyer, P. Ogrocki, and C. E. Speicher, "Chronic stress and immunity in family caregivers of Alzheimer's disease victims," *Psychosomatic Medicine* 49 (1987): 523–535. J. K. Kiecolt-Glaser, L. D. Fisher, P. Ogrocki, J. C. Stout, C. E. Speicher, and R. Glaser, "Marital quality, marital disruption, and immune function," *Psychosomatic Medicine* 49 (1987): 13–34. J. K. Kiecolt-Glaser, S. Kennedy, S. Malkoff, L. Fisher, C. E. Speicher, and R. Glaser, "Marital discord and immunity in males," *Psychosomatic Medicine* 50 (1988): 213–229.

The two studies by Drs. Glaser and Kiecolt-Glaser demonstrating beneficial effects of stress reduction and enhancement of positive emotions described in this chapter are: J. K. Kiecolt-Glaser, R. Glaser, D. Williger, J. Stout, G. Messick, S. Sheppard, D. Ricker, S. C. Romisher, W. Briner, G. Bonnell, and R. Donnerberg, "Psychosocial enhancement of immunocompetence in a geriatric population," *Health Psychology* 4 (1985): 24–41. J. K. Kiecolt-Glaser, R. Glaser, E. C. Strain, J. C. Stout, K. L. Tarr, J. E. Holliday, and C. E. Speicher, "Modulation of cellular immunity in medical students," *Journal of Behavioral Medicine* 9 (1986): 5–21.

The ordeal of the medical internship is discussed in an article by Norman Cousins, "Internship: Preparation or Hazing?" *Journal of the American Medical Association* 245 (1981): 377.

Norman Cousins reports on current findings on the way in which emotions, experiences, and attitudes can create physiological change in the following article: "Intangibles in Medicine: An Attempt at a Balancing Perspective," *Journal of the American Medical Association* 260 (1988): 1610–1612.

4. The Patient-Physician Relationship

The Physicians' Desk Reference is published annually by the Medical Economics Company, Inc., in Oradell, New Jersey.

Statistics on the use of Valium can be found in Eve Bargmann, Sidney M. Wolfe, and Joan Levin, *Stopping Valium, Ativan, Centrax, Dalmane, Librium, Paxipam, Restoril, Serax, Tranxene & Xanax* (Washington, D.C.: Public Citizen's Health Research Group, 1982): 4.

Dr. Henri Manasse has produced a report on drug misadventuring that will appear in a series of articles to be published in the *American Journal of Hospital Pharmacy* and will later be condensed into a book, *Medication Use in an Imperfect World: Drug Misadventuring as an Issue of Public Policy.*

Many references on medication-related hazards, including the overuse of antibiotics, are cited in Charles B. Inlander, Lowell S. Levin, and Ed Weiner, *Medicine on Trial* (New York: Prentice-Hall, 1988), 135–153.

Additional information on the overuse of antibiotics can be found in the following publications: A. W. Roberts and J. A. Visconti, "The rational and irrational use of systemic antimicrobial drugs," *American Journal of Hospital Pharmacy* 29 (1972): 828–834. H. E. Simmons and P. D. Stolley, "This is medical progress? Trends and consequences of antibiotic use in the United States," *Journal of the American Medical Association* 227 (1974): 1023–1028.

Information on the factors contributing to the increasing use of non–medically justified drug prescribing (including tranquilizers, antibiotics, and other drugs) and possible solutions to the problem can be found in Ingrid Waldron, "Increased prescribing of Valium, Librium, and other drugs—an example of the influence of economic and social factors on the practice of medicine," *International Journal of Health Services* 71 (1977): 37–62.

The doctor-patient relationship is discussed in an article by Norman Cousins, "Unacceptable Pressures on the Physician," *Journal of the American Medical Association* 252 (1984): 351–352.

5. The Infinite Wonder of the Human Brain (And How It Affects Even AIDS)

Dr. Clemente's widely used textbook on anatomy is *Anatomy: A Regional Atlas of the Human Body*, 3rd edition (Baltimore: Urban and Schwarzenberg, 1987).

Carmine D. Clemente, editor, *Gray's Anatomy of the Human Body*, 30th American edition (Philadelphia: Lea & Feibiger, 1985).

A detailed account of the history of the Brain Research Institute is recorded in John D. French, Donald B. Lindsley, and H. W. Magoun, *An American Contribution to Neuroscience: The Brain Research Institute, UCLA* 1959–1984 (Los Angeles: UCLA Publication Services Department, 1984).

Drs. Elena A. Korneva and Viktor M. Klimenko's research on neuroimmune interactions is summarized in E. A. Korneva, V. M. Klimenko, and E. R. Shkhinek, *Neurohumoral Maintenance of Immune Homeostasis*, translated from the Russian by Samuel A. and Elizabeth O'Leary Corson in collaboration with Roland Dartau, Justina Epp, and L. A. Mutschler (Chicago: University of Chicago Press, 1985).

Dr. Nicholas P. Plotnikoff's experiments on the use of enkephalins in the alteration of immunity is reported in N. P. Plotnikoff, A. Murgo, R. E. Faith, and R. Good, *Enkephalins and Endorphins: Stress and the Immune System* (New York: Plenum Press, 1986).

Dr. Candace B. Pert's discussion of the role of neuropeptides as a communicating link between mind and body was published in her article, "The wisdom of the receptors: Neuropeptides, the emotions, and bodymind," *Advances* 3 (1986): 8–16. A more technical presentation of this information can be found in C. B. Pert, M. R. Ruff, R. J. Weber, and M. Herkenham, "Neuropeptides and their receptors: A psychosomatic network," *Journal of Immunology* 135 (1985): 820S–826S.

The role of neuropeptides in the functioning of the immune system is explored in J. E. Morley, N. E. Kay, G. F. Solomon, and N. P. Plotnikoff, "Neuropeptides: Conductors of the immune orchestra," *Life Sciences*, 41 (1987): 527–544.

Dr. George Solomon's observations of the immune profiles of long-survivors of AIDS is in preparation.

6. Denial, No. Defiance, Yes.

More information about the Wellness Community can be obtained by writing to: The Wellness Community, 1235 Fifth Street, Santa Monica, CA 90401.

Drs. Sandra M. Levy and Ronald B. Herberman's research on the connection between passivity and depression, and immune activity in cancer patients is reported in S. M. Levy, R. B. Herberman, A. M. Maluish, B. Schlein, and M. Lippman, "Prognostic risk assessment in primary breast cancer by behavioral and immunological parameters," *Health Psychology* 4 (1985): 99–113.

A series of studies demonstrating immunological responses to depression conducted by Drs. Marvin Stein, Steven Schleifer, and Steven Keller were published in the following journals: S. J. Schleifer, S. E. Keller, M. Camerino, J. C. Thornton, and M. Stein, "Suppression of lymphocyte stimulation following bereavement," *Journal of the American Medical Association* 250 (1983): 374–377. S. J. Schleifer, S. E. Keller, A. T. Meyerson, M. J. Raskin, K. L. Davis, and M. Stein, "Lymphocyte function in major depressive disorder," *Archives of General Psychiatry* 41 (1984): 484–486. S. J. Schleifer, S. E. Keller, S. G. Siris, K. L. Davis, and M. Stein, "Depression and immunity: Lymphocyte function in ambulatory depressed, hospitalized schizophrenic, and herniorrhaphy patients," *Archives of General Psychiatry* 42 (1985): 129–133. M. Stein, S. E. Keller, and S. J. Schleifer, "Stress and immunomodulation: The role of depression and neuroendocrine function," *The Journal of Immunology* 135 (1985): 827–833.

The research by Drs. Ronald Glaser and Janice Kiecolt-Glaser regarding suboptimal DNA repair in lymphocytes following irradiation in depressed patients can be found in J. K. Kiecolt-Glaser, R. E. Stephens, P. D. Lipetz, C. E. Speicher, and R. Glaser, "Distress and DNA repair in human lymphocytes," *Journal of Behavioral Medicine* 8 (1985): 311–320.

Drs. Barry L. Gruber and Nicholas R. S. Hall's project involving the use of relaxation and guided imagery with cancer patients is in B. L. Gruber, N. R. Hall, S. P. Hersh, and P. Dubois, "Immune system and psychologic changes in metastatic cancer patients while using ritualized relaxation and guided imagery: A pilot study," *Scandinavian Journal of Behavior Therapy* 17 (1988): 25–46.

The scientific evidence that depression can inhibit the effectiveness of the immune system is summarized by Joseph R. Calabrese, Mitchel A. Kling, and Philip W. Gold of the National Institute of Mental Health in Rockville, Maryland, in "Alterations in immunocompetence during stress, bereavement, and depression: Focus on neuroendocrine regulation," *American Journal of Psychiatry* 144 (1987): 1123–1134. The article also describes components of the immune system, implications of immune abnormalities, and possible biochemical factors that produce immune changes.

A discussion of ways to evaluate denial in patients appears in an article by Norman Cousins, "Denial," *Journal of the American Medical Association* 248 (1982): 210–212.

7. False Hopes vs. False Fears

The quotation by Dr. William M. Buchholz regarding the medical uses of hope appeared in the *Western Journal of Medicine* 148 (1988): 69.

8. Illness and Guilt

Dr. Bernie S. Siegel's goals as a physician are described in his book, *Love, Medicine & Miracles* (New York: Harper and Row, 1986).

9. Problems Beyond the Doctor's Reach

Dr. Franz Ingelfinger's statement that 85 percent of human illnesses are self-limiting was printed in his article "Arrogance," *New England Journal of Medicine* 303 (1980): 1506–1511.

10. The Laughter Connection

The Swedish study demonstrating the use of humor to improve the quality of life and symptom relief appeared in Lars Ljungdahl, "Laugh if this is a joke," *Journal of the American Medical Association* 261 (1989): 558.

Norman Cousins' first account of his triumph over illness, "Anatomy of an Illness," was published in the *New England Journal of Medicine* 295 (1976): 1458–1463. The book *Anatomy of an Illness* (New York: W. W. Norton, 1979) is a fuller version of that article.

Dr. James J. Walsh reported his seminal observations on the effects of laughter on internal organs in *Laughter and Health* (New York: D. Appleton and Company, 1928).

Dr. William F. Fry, Jr.'s description of laughter as a form of exercise appeared in the Health Briefing Section of *Insight*, May 25, 1987, 59.

The physiology of humor has been synopsized by Dr. William F. Fry, Jr., in "Humor, Physiology, and the Aging Process," in Lucille Nahemow, Kathleen A. McCluskey-Fawcett, and Paul E. McGhee, Eds., *Humor and Aging* (Orlando: Academic Press, 1986), 81–98.

Dr. Gordon Allport's view of humor as contributing to a fresh perspective in life is articulated in *The Individual and His Religion* (New York: Macmillan, 1950).

Dr. Annette Goodheart's belief in the contribution of humor toward a more creative perspective is reported in J. M. Leighty, "Laughter Helps the Heart and Soul," *The Houston Chronicle*, June 9, 1987.

Dr. Alice M. Isen's research showing the creativity-inducing characteristics of laughter were reported in the following publications: A. M. Isen, M. M. S. Johnson, E. Mertz, and G. F. Robinson, "The influence of positive affect on the unusualness of word associations," *Journal of Personality and Social Psychology* 48 (1985): 1413–1426. A. M. Isen,

K. A. Daubman, and G. P. Nowicki, "Positive affect facilitates creative problem solving," *Journal of Personality and Social Psychology* 52 (1987): 1122–1131. A. M. Isen, "Positive affect, cognitive processes, and social behavior," *Advances in Experimental Social Psychology* 20 (1987): 203–253.

Drs. Rosemary and Dennis Cogan's study demonstrating the efficacy of laughter in reducing "discomfort sensitivity" is reported in R. Cogan, D. Cogan, W. Waltz, and M. McCue, "Effects of laughter and relaxation on discomfort thresholds," *Journal of Behavioral Medicine* 10 (1987): 139–144.

Drs. Rod A. Martin and Herbert M. Lefcourt's series of studies exploring the use of humor in counteracting the negative emotional effects of stress appeared in R. A. Martin and H. M. Lefcourt, "Sense of humor as a moderator of the relation between stressors and moods," *Journal of Personality and Social Psychology* 45 (1983): 1313–1324.

Dr. James R. Averill's experiment demonstrating the physiological manifestations of different emotions was reported in J. R. Averill, "Autonomic response patterns during sadness and mirth," *Psychophysiology* 5 (1969): 399–414.

Dr. Paul Eckman's research determining the physiological characteristics of various emotional states was published in P. Eckman, R. W. Levenson, and W. V. Friesen, "Autonomic nervous system activity distinguishes among emotions," *Science*, 221 (1983): 1208–1210.

Norman Cousins, *Anatomy of an Illness* (New York: W. W. Norton, 1979).

Dr. Gary E. Schwartz's study of the physiological effects of different emotions during exercise was printed in G. E. Schwartz, D. A. Weinberger, and J. A. Singer, "Cardiovascular differentiation of happiness, sadness, anger, and fear following imagery and exercise," *Psychosomatic Medicine* 43 (1981): 343–364.

Dr. Kathleen M. Dillon's projects demonstrating the positive effects of laughter on immunity appear in the following publications: K. M. Dillon, B. Minchoff, and K. H. Baker, "Positive emotional states and enhancement of the immune system," *International Journal of Psychiatry in Medicine* 15 (1985–86): 13–17. K. M. Dillon and M. C. Totten, "Psychological factors affecting immunocompetence and health of breastfeeding mothers and their infants," *Journal of Genetic Psychology* 150 (1989). Forthcoming.

Dr. David C. McClelland's research linking positive changes in immunity to positive emotional states is reviewed in the following articles: D. C. McClelland, "Some reflections on the two psychologies of love," *Journal of Personality* 54 (1986): 334–353. D. C. McClelland

and C. Kirshnit, "The effect of motivational arousal through films on salivary immunoglobulin A," *Psychology and Health* 2 (1988): 31–52.

Dr. Lee S. Berk's studies of the immune-enhancing effects of laughter have been described in the following publications: L. S. Berk, S. A. Tan, S. L. Nehlsen-Cannarella, B. J. Napier, J. W. Lee, J. E. Lewis, R. W. Hubbard, and W. C. Eby, "Mirth modulates adrenocorticomedullary activity: Suppression of cortisol and epinephrine," *Clinical Research* 36 (1988): 121. A. L. S. Berk, S. A. Tan, S. L. Nehlsen-Cannarella, B. J. Napier, J. E. Lewis, J. W. Lee, W. F. Fry, and W. C. Eby, "Laughter decreases cortisol, epinephrine and 3,4 dihydroxyphenyl acetic acid (Dopac)," *The Federation of American Societies for Experimental Biology (FASEB) Journal* 2 (1988): A1570. L. S. Berk, S. A. Tan, S. L. Nehlsen-Cannarella, B. J. Napier, J. E. Lewis, J. W. Lee, and W. C. Eby, "Humor associated laughter decreases cortisol and increases spontaneous lymphocyte blastogenesis," *Clinical Research* 36 (1988): 435A.

A study of the immune-suppressing effects of corticosteroid therapy which corroborates Dr. Berk's findings appears in P. Pinkston, C. Saltini, J. Müller-Quernheim, and R. G. Crystal, "Corticosteroid therapy suppresses spontaneous interleukin 2 release and spontaneous proliferation of lung T lymphocytes of patients with active pulmonary sarcoidosis," *The Journal of Immunology* 139 (1987): 755–760.

Jane Brody explores the use of humor by health care professionals in the treatment of patients in "Personal Health: Increasingly, laughter as potential therapy for patients is being taken seriously," *The New York Times*, April 7, 1988, B8.

Kaye Ann Herth's humor history-taking and the humor handbook of the Andrus Gerontology Center at the University of Southern California are described in the *American Association of Retired Persons (AARP) News Bulletin*, June 1987, 1 and 10. The AARP is based in Washington, D.C.

11. Medical Mistakes

Edward C. Lambert, *Modern Medical Mistakes* (Bloomington: Indiana University Press, 1978).

Statistics regarding the number of bypass operations performed in the mid-1980s can be found in *1988 Heart Facts*, an annual publication of the American Heart Association. Copies can be obtained by calling the Heart Information Service of the American Heart Association (1-800-432-7852).

The National Heart, Lung, and Blood Institute study of bypass operations entitled, "Myocardial infarction and mortality in the coronary artery surgery study (CASS) randomized trial," was published in *New England Journal of Medicine* 310 (1984): 750–758.

The corroborating results of the European bypass study were reported in E. Varnauskas, "Myocardial infarction in the randomized European coronary surgery study," *Circulation* 68 (1983), supplement 3: III-293.

Historical treatments of bone fractures are reviewed in Lambert, *Modern Medical Mistakes*.

The 1976 congressional subcommittee on Oversight and Investigations of the House Committee on Interstate and Foreign Commerce findings on surgeries lacking sufficient justification were reviewed in *Unnecessary Surgery: Double Jeopardy for Older Americans*, a hearing before the U.S. Senate Special Committee on Aging, published by the U.S. Government Printing Office, Washington, D.C., 1985.

The National Institute of Health Statistics report on hysterectomies was cited in Charles B. Inlander, Lowell S. Levin, and Ed Weiner, *Medicine on Trial* (New York: Prentice-Hall, 1988), 114.

The UCLA and Rand Corporation studies of the appropriateness of commonly used medical procedures were reported in the following publications: C. M. Winslow, D. H. Solomon, M. R. Chassin, J. Kosecoff, N. J. Merrick, and R. H. Brook, "The appropriateness of carotid endarterectomy," *New England Journal of Medicine* 318 (1988): 721–727. C. M. Winslow, J. B. Kosecoff, M. Chassin, D. E. Kanouse, and R. H. Brook, "The appropriateness of performing coronary artery bypass surgery," *Journal of the American Medical Association* 260 (1988): 505–509. M. R. Chassin, J. Kosecoff, D. H. Solomon, and R. H. Brook, "How coronary angiography is used: Clinical determinants of appropriateness," *Journal of the American Medical Association* 258 (1987): 2543–2547. M. R. Chassin, J. Kosecoff, R. E. Park, C. M. Winslow, K. L. Kahn, N. J. Merrick, J. Keesey, A. Fink, D. H. Solomon, and R. H. Brook, "Does inappropriate use explain geographic variations in the use of health care services? A study of three procedures," *Journal of the American Medical Association* 258 (1987): 2533–2537.

The study of unwarranted cardiac pacemaker implantations appears in A. M. Greenspan, H. R. Kay, B. C. Berger, R. M. Greenberg, A. J. Greenspon, and M. J. S. Gaughan, "Incidence of unwarranted implantation of permanent cardiac pacemakers in a large medical population," *New England Journal of Medicine* 318 (1988): 158–163.

Dangers in the usage of X rays are summarized in John W. Gofman and Egan O'Connor, *X-Rays: Health Effects of Common Exams*

(San Francisco: Sierra Club Books, 1985), and is cited in Inlander et al., *Medicine on Trial* (New York: Prentice Hall Press, 1988): 106–108.

Dr. Manfred Sakel's observations of the effects of insulin injections on schizophrenic patients is described in Lambert, *Modern Medical Mistakes*.

The Johns Hopkins University Hospital investigation of the connection between the prenatal use of hormones and children's deformities is described in Lambert, *Modern Medical Mistakes*.

The Physicians' Desk Reference is published annually by the Medical Economics Company, Inc., in Oradell, New Jersey.

Eric W. Martins with Arthur Ruskin, medical editor, and Ruth Martin, associate editor, *Hazards of Medication: A Manual on Drug Interactions, Contraindications, and Adverse Reactions, with Other Prescribing and Drug Information* (Philadelphia: Lippincott, 1978).

12. Mesmer, Hypnotism, and the Powers of Mind

Dr. Richard Bergland has written about secretions of the brain in his book, *The Fabric of Mind* (Middlesex, England: Viking Penguin, 1985).

The dangers of hypnotism have been outlined in Louis J. West and Gordon H. Deckert, "Dangers of hypnosis," *Journal of the American Medical Association* 192 (1965): 9–12.

James Braid's role in the labeling and documentation of the phenomenon of hypnotism is described in André M. Weitzenhoffer, *Hypnotism An Objective Study in Suggestibility* (New York: John Wiley and Sons, 1953).

The following resources on the life and contribution of Franz Anton Mesmer were compiled by Dr. Dora B. Weiner of UCLA: R. Darnton, *Mesmerism and the End of Enlightenment in France* (Cambridge: Harvard University Press, 1968). Henri F. Ellenberger, *The Discovery of the Unconscious: The History and Evolution of Dynamic Psychiatry* (New York: Basic Books, 1970), 53–253. C. G. Gillispie, *"Mesmerism" in Science and Polity in France at the End of the Old Regime* (Princeton, N.J.: Princeton University Press, 1980), 261–289. F. A. Mesmer, *Mesmerism: A Translation of the Original Scientific and Medical Writings of F. A. Mesmer*, translated by George Bloch, with introduction by E. R. Hilgard (Los Altos, Calif.: William Kaufmann, 1980). G. Sutton, "Electric medicine and Mesmerism," *Isis* 72 (1981): 375–392.

14. The Physician in Literature

Sir William Osler, *Aequanimitas*, 3rd edition (New York: Blakiston Company, 1953).

The world of scientists and humanists described by Charles Percy (C. P.) Snow is explored in his book *The Two Cultures* (Cambridge: Cambridge University Press, 1979).

Most of the works of literature referred to throughout this chapter are cited in *The Physician in Literature*, edited, with an introduction, by Norman Cousins (Philadelphia: Saunders Press, 1982).

15. Task Force Beginnings

Dr. Hans Selye describes the process of scientific discovery in his book *From Dream to Discovery* (Salem, N.H.: Ayer Company, 1964).

Hans Zinsser, *As I Remember Him: The Biography of R. S.* (Boston: Little, Brown, 1940).

Mr. Brendan O'Regan's collection of reports on spontaneous remissions of immune-related disorders is in preparation.

Dr. Franz Ingelfinger's article on the self-limiting characteristic of ailments appears in his article "Arrogance," *New England Journal of Medicine* 303 (1980): 1506–1511.

Dr. John Liebeskind's research on the body's production of opioid and nonopioid analgesia is described in J. C. Liebeskind, J. W. Lewis, Y. Shavit, G. W. Terman, and T. Melnechuk, "Our natural capacities for pain suppression," *Advances* 1 (1983): 8–11. A more technical discussion of intrinsic analgesia can be found in J. C. Liebeskind, J. E. Sherman, J. T. Cannon, and G. W. Terman, "Neural and neurochemical mechanisms of pain inhibition," *Seminars in Anesthesia* 4 (1985): 218–222.

16. On the Defensive

The controversial article by Dr. Barrie R. Cassileth and colleagues, referred to throughout this chapter, appeared in B. R. Cassileth, E. J. Lusk, D. S. Miller, L. L. Brown, and C. Miller, "Psychosocial correlates of survival in advanced malignant disease?" *New England Journal of Medicine* 312 (1985): 368–373. The editorial by Dr. Marcia Angell, "Disease as a reflection of the psyche," follows the Cassileth article on pages 373–375.

"Anatomy of an Illness," written by Norman Cousins, first appeared in the *New England Journal of Medicine* 295 (1976): 1458–1463.

The book *Anatomy of an Illness* (New York: W. W. Norton, 1979) is a fuller version of that article.

A general reference is made in this chapter to research identifying psychological factors associated with the *onset* of cancer. For the reader's information, a review of the history of research on the cancer-prone personality profile can be found in: C. B. Bahnson, "Psychological aspects of cancer," in *Surgical Oncology* (New York: McGraw-Hill, 1984): 231–253. C. B. Bahnson, "Stress and cancer: The state of the art," *Psychosomatics*, Part I, 21 (1980): 975–981, and Part II, 22 (1981): 207–220. S. Greer and M. Watson, "Towards a psychobiological model of cancer: Psychological considerations," *Social Science and Medicine* 20 (1985): 773–777.

Also for the reader's information, long-term studies predicting the *onset* of cancer on the basis of psychological and life history factors are as follows: A. Bremond, G. Kune, and C. B. Bahnson, "Psychosomatic factors in breast cancer patients: Results of a case controlled study," *Journal of Psychosomatic Obstetrics and Gynecology*, 5 (1986): 127–136. R. Grossarth-Maticek, "Psychosocial predictors of cancer and internal diseases: An overview," *Psychotherapy and Psychosomatics* 33 (1980): 122–128. R. Grossarth-Maticek, J. Bastiaans, and D. T. Kanazir, "Psychosocial factors as strong predictors of mortality from cancer, ischaemic heart disease and stroke: The Yugoslav prospective study," *Journal of Psychosomatic Research*, 29 (1985): 167–176. M. Harrower, C. B. Thomas, and A. Altman, "Human figure drawings in a prospective study of six disorders: Hypertension, coronary heart disease, malignant tumor, suicide, mental illness, and emotional disturbance," *Journal of Nervous and Mental Disease* 161 (1975): 191–199. R. L. Horne and R. S. Picard, "Psychosocial risk factors for lung cancer," *Psychosomatic Medicine* 41 (1979): 503–514. A. H. Schmale and H. P. Iker, "Hopelessness as a predictor of cervical cancer," *Social Science and Medicine* 5 (1971): 95. C. B. Thomas, "Precursors of premature disease and death: The predictive potential of habits and family attitudes," *Annals of Internal Medicine* 85 (1976): 653–658. C. B. Thomas, K. R. Duszynski, and J. W. Shaffer, "Family attitudes reported in youth as potential predictors of cancer," *Psychosomatic Medicine* 41 (1979): 287–302.

The long-term studies described in this chapter predicting the *course* of cancer on the basis of psychological characteristics were reported in the following publications: L. R. Derogatis, M. D. Abeloff, and N. Melisaratos, "Psychological coping mechanisms and survival time in metastic breast cancer," *Journal of the American Medical Association* 242 (1979): 1504–1508. H. S. Greer, T. Morris, and K. W. Pettingale, "Psychological response to breast cancer: Effect on outcome," *Lancet* 2 (1979): 785–787. K. W. Pettingale, T. Morris, S. Greer, and J. S. Haybittle, "Mental attitudes to cancer: An additional prognostic factor," *Lancet* 8 (1985): 750. S. M. Levy, R. B. Herberman, A. M. Maluish, B. Schlein, and M. Lippman, "Prognostic risk assessment in primary breast cancer by behavioral and immunological parameters," *Health Psychology* 4 (1985): 99–113. S. Levy, R. Herberman, M. Lippman, and T. d'Angelo, "Correlation of stress factors with sustained depression of natural killer cell activity and predicted prognosis in patients with breast cancer," *Journal of Clinical Oncology* 5 (1987): 348–353. G. N. Rogentine, Jr., D. P. Van Kammen, B. H. Fox, J. P. Docherty, J. E. Rosenblatt, S. C. Boyd, and W. E. Bunney, "Psychological factors in the prognosis of malignant melanoma: A prospective study," *Psychosomatic Medicine* 41 (1979): 647–655. L. Temoshok, "Biopsychosocial studies on cutaneous malignant melanoma: Psychological factors associated with prognostic indicators, progression, psychophysiology, and tumor-host response," *Social Science and Medicine* 20 (1985): 833–840.

The UCLA survey of oncologists regarding the importance of the role of psychosocial factors in the health of their patients is in preparation.

18. Belief Becomes Biology

Dr. Ronald Katz's observation regarding spinal anesthesia headaches was reported in his article "Informed consent—Is it bad medicine?" *The Western Journal of Medicine* 126 (1977): 426–428, and via personal communication on August 29, 1988.

The study of hair loss as a result of expectations attached to a placebo was reported by J. W. L. Fielding, S. L. Fagg, B. G. Jones, S. Ellis, M. S. Hockey, A. Minawa, V. S. Brookes, J. L. Craven, M. C. Mason, A. Timothy, J. A. H. Waterhouse, and P. F. M. Wrigley, "An interim report of a prospective, randomized, controlled study of adjuvant chemotherapy in operable gastric cancer: British stomach cancer group," *World Journal of Surgery* (1983): 390–399.

Dr. Neal E. Miller's research on placebo phenomena and its implications is outlined in his article "Placebo factors in treatment: Views of a psychologist," in M. Shepherd and N. Sartorius, editors, *Non-Specific Aspects of Treatment* (Bern, Stuttgart, and Toronto: Hans Huber Verlag, 1989).

Drs. Robert Ader and Anthony Suchman's work on conditioning the body to respond to placebos is summarized in R. Ader and A. Suchman, "CNS-immune system interactions: Conditioning phenomena," *Behavioral and Brain Sciences* 8 (1985): 379–426.

Drs. S. Metal'nikov and V. Chorine reported the first accounts of immune conditioning phenomenon in their article "The role of conditioned reflexes in immunity," *Annals of the Pasteur Institute* 40 (1926): 893–900.

Dr. Robert Ader explains the principles of conditioning and implications of conditioning research on the evaluation and application of medication therapy in his article "Conditioning effects in pharmacotherapy and the incompleteness of the double-blind, crossover design," *Integrative Psychiatry*. Forthcoming.

Drs. Ader and Suchman's report on the effectiveness of conditioning techniques in the management of hypertension is in preparation.

Dr. Henry K. Beecher described his observation that more than a third of all patients experience significant placebo effects in his article "The powerful placebo," *Journal of the American Medical Association* 159 (1955) 1602–1606.

Some of the physiological reactions that have been shown to be responsive to placebos are listed in S. Ross and L. Buckalew, "The placebo as an agent of behavioral manipulation: A review of problems, issues and affected measures," *Clinical Psychology Review* 3 (1983): 457–471.

The study by Dr. Jon D. Levine demonstrating that endorphins play a role in postoperative dental pain relief from a placebo is published in J. D. Levine, N. C. Gordon, H. L. Fields, and L. Howard, "The mechanism of placebo analgesia," *Lancet* 2 (1978): 654–657.

A discussion of the implications of placebo analgesia for medical treatment can be found in J. C. Liebeskind, J. W. Lewis, Y. Shavit, G. W. Terman, and T. Melnechuk, "Our natural capacities for pain suppression," *Advances* 1 (1983): 11.

A summary of the issues in and implications of placebo research, as well as a discussion of physiological mechanisms involved in placebo effects is contained in *Investigations* 2 (1985). *Investigations* is a research bulletin of the Institute of Noetic Sciences, a non-profit foundation located at 475 Gate Five Road, Suite 300, Sausalito, CA 94965. This particular issue is entitled "Placebo—The hidden asset in healing."

Any projects supported by UCLA that are not referenced below are in preparation and have not been published yet, or are educational in nature and do not involve a publication:

HERBERT BENSON

Although Dr. Herbert Benson's research project is currently in preparation, there are two references that describe the phenomenon and application of the relaxation response to the treatment of patients:

J. W. Hoffman, H. Benson, P. A. Arns, G. L. Stainbrook, L. Landsberg, J. B. Young, and A. Gill "Reduced sympathetic nervous system responsivity associated with the relaxation response," *Science* 215 (1982): 190–192. G. L. Stainbrook, J. W. Hoffman, and H. Benson, "Behavioral therapies of hypertension: Psychotherapy, biofeedback, and relaxation/meditation," *International Review of Applied Psychology* 32 (1983): 119–135.

MICHAEL IRWIN

M. Irwin, R. L. Hauger, M. Brown, and K. T. Britton, "CRF activates autonomic nervous system and reduces natural killer cytotoxicity," *American Journal of Physiology* 255 (November 1988): R744–R747.

SHERRIE H. KAPLAN AND SHELDON GREENFIELD

S. H. Kaplan, S. Greenfield, and J. E. Ware, Jr., "Assessing the effects of physician-patient interactions on the outcomes of chronic disease," *Journal of Medical Care* 27(Supplement) (1989): S110–S127.

ALFRED H. KATZ

A. H. Katz, C. A. Maida, G. Strauss, and C. Kwa, "Social adaptation in a chronic disease: A study of lupus," in Kutscher Austin H., editor, *Self-Help Group: Life-and Personhood-Threatening Conditions* (Philadelphia: Charles Press, 1989).

STEVEN E. LOCKE

S. E. Locke and M. Hornig-Rohan, *Mind and Immunity: Behavioral Immunology* (New York: Institute for the Advancement of Health, 1983). Contains 1,453 research abstracts and references.

S. E. Locke, B. J. Ransil, N. A. Covino, J. Toczydlowski, C. M. Lohse, H. F. Dvorak, K. A. Arndt, and F. H. Frankel, "Failure of hypnotic suggestion to alter immune response to delayed-type hypersensitivity antigens," *Annals of the New York Academy of Sciences* 496 (1987): 745–749.

THEODORE MELNECHUK

Although Mr. Melnechuk's report is in preparation, the article listed here provides a synopsis of the forthcoming publication: T. Melnechuk, "Emotions, Brain, Immunity, and Health: A Review," in M. Clynes and J. Panksepp, editors, *Emotions and Psychopathology* (New York: Plenum Press, 1988), 181–247. Contains 530 references.

JOSEPHINE T. RHODES

J. Rhodes, "A controlled trial of the effects of professional peer group counseling in rheumatoid arthritis" in Paul Ahmed, editor, *Coping with Arthritis* (Springfield, Ill.: Charles E. Thomas, 1988), 73–106. J. Rhodes, "Rheumatoid Arthritis—A Dialogue with Pain: From Subjective Experience to Objective Observation," in Ahmed, editor, *Coping with Arthritis* 107–125.

ALAN ROZANSKI

L. Yang, C. N. Bairey, A. Rozanski, K. Nichols, J. Friedman, J. Areeda, K. Suyenaga, N. Syun, and D. Berman, "Validation of the ambulatory function monitor (VEST) for measuring exercise left ventricular ejection fraction," *Journal of Nuclear Medicine* 29 (1988): 741. A. Rozanski, C. N. Bairey, D. S. Krantz, J. Friedman, K. J. Resser, M. Morell, S. Hilton-Chalfen, L. Hestrin, J. Bietendorf, and D. S. Berman, "Mental stress in the induction of silent myocardial ischemia in patients with coronary artery disease," *The New England Journal of Medicine* 318 (1988): 1005–1012. P. LaVeau, A. Rozanski, D. Krantz, C. Cornell, L. Cattanach, B. L. Zaret, and F. Wackers, "Ischemic left ventricular performance during provocative mental stress testing," *Circulation* 74 (1986): II-504.

GEORGE F. SOLOMON

G. F. Solomon, M. A. Fiatarone, D. Benton, J. E. Morley, E. Bloom, and T. Makinodan, "Psychoimmunologic and endorphin function in the aged," *Annals of the New York Academy of Sciences* 521 (1988): 43–58. M. A. Fiatarone, J. E. Morley, E. T. Bloom, D. Benton, G. F. Solomon, and T. Makinodan, "The effect of exercise on natural killer cell activity in young and old subjects," *Journal of Gerontology*. In press. G. F. Solomon, L. Temoshok, A. O'Leary, and J. Zich, "An intensive psychoimmunologic study of long-surviving persons with AIDS," *Annals of the New York Academy of Sciences* 496 (1987): 647–655. The report of the follow-up investigation of the relationship between psychosocial factors and immune measures in AIDS patients is in preparation.

The study corroborating Dr. Solomon's finding that age plays a significant role in the interface between emotions and immunity is published in S. J. Schleifer, S. E. Keller, R. N. Bond, J. Cohen, and M. Stein, "Major depressive disorder and immunity," *Archives of General Psychiatry* 46 (1989): 81–87.

19. Functional Age

The excerpt from Drs. George F. Solomon and John E. Morley's report on the immune and psychological factors contributing to the health of elderly persons was taken from G. F. Solomon, M. A. Fiatarone, D. Benton, J. E. Morley, E. Bloom, and T. Makinodan, "Psychoimmunologic and endorphin function in the aged," *Annals of the New York Academy of Sciences* 521 (1988): 43–58.

Dr. Elie Metchinkoff, *The Nature of Man: Studies in Optimistic Philosophy* (English translation) was edited by P. Chalmers Mitchell and published by Heinemann, London, 1906.

Dr. Nathan Shock's research on the extension of the life span through the conquest of disease can be found in the following publications:

N. W. Shock, "The physiological basis of aging," in Robert J. Morin and Richard J. Bing (eds.), *Frontiers in Medicine: Implications for the Future* (New York: Human Sciences Press, 1985), 300–312. N. W. Shock, "Longitudinal studies of aging in humans," in C. Finch and E. Schneider (eds.), *Handbook of the Biology of Aging* (New York: Van Nostrand Reinhold, 1985), 721–743. N. W. Shock, "The evolution of gerontology as a science," in Morton Rothstein (ed.), *Review of Biological Research on Aging*, vol. 3 (New York: Alan R. Liss, 1987), 3–12.

20. The Society of Challengers

and

21. Obsession Revisited

The study of the psychosocial enhancement of patients with malignant melanoma by Dr. Fawzy I. Fawzy, Dr. John L. Fahey, Dr. Donald L. Morton, and Mr. Norman Cousins is in preparation.

22. Report to the Dean

A number of the investigators who made presentations at the UCLA conference on psychoneuroimmunology are referenced in other chapters. Herewith, a listing of publications by some of the individuals whose work described in this chapter is not referenced elsewhere which may be of interest to the reader:

RUDY E. BALLIEUX

C. J. Heijnen and R. E. Ballieux, "Influence of opioid peptides on the immune system," *Advances* 3 (1986): 114–121. G. Croiset, C. J. Heijnen, H. D. Veldhuis, D. de Wied, and R. E. Ballieux, "Modulation of the immune response by emotional stress," *Life Sciences* 40 (1987): 775–782. G. Croiset, H. D. Veldhuis, R. E. Ballieux, D. de Wied, and C. J. Heijnen, "The impact of mild emotional stress induced by the passive avoidance procedure on immune reactivity," *Annals of the New York Academy of Sciences* 496 (1987): 477–484. C. J. Heijnen, J. Zijlstra, A. Kavelaars, G. Croiset, and R. E. Ballieux, "Modulation of the immune response by POMC-derived peptides. I. Influence on proliferation of human lymphocytes," *Brain, Behavior, and Immunity* 1 (1987): 284–291.

HUGO O. BESEDOVSKY

H. Besedovsky, A. del Rey, E. Sorkin, and C. A. Dinarello, "Immunoregulatory feedback between interleukin-1 and glucocorticoid hormones," *Science* 233 (1986): 652–654. H. Besedovsky and A. del Rey, "Neuroendocrine and metabolic responses induced by interleukin-1," *Journal of Neuroscience Research* 18 (1987): 172–178. F. Berkenbosch, A. del Rey, J. Van Oers, F. Tilders, and H. Besedovsky, "Feedback circuit involving the immunohormone interleukin-1 and the hypothalamus-pituitary adrenal system," in R. Kvetnansky, editor, *Catecholamines and Other Neurotransmitters in Stress* (New York: Gordon and Breach, 1988).

KAREN BULLOCH

K. Bulloch, "Neuroanatomy of lymphoid tissue: A review," in R. G. Guillemin, M. Cohen, and T. Melnechuk, editors, *Neural Modulation of Immunity* (New York: Raven Press, 1985), 111–141. K. Bulloch, M. R. Cullen, R. H. Schwartz, and D. L. Longo, "Development of innervation within syngeneic thymus tissue transplanted under the kidney capsule of the nude mouse: A light and ultrastructural microscope study," *Journal of Neuroscience Research* 18 (1987): 16–27. This article also appears in *Neuroimmunomodulation*, edited by J. R. Perez-Polo, K. Bulloch, R. H. Angeletti, G. A. Hashim, and J. de Vellis (New York: Alan R. Liss, 1987). K. Bulloch and R. Lucito, "The effects of cortisone on acetylcholinesterase (AChE) in the neonatal and aged thymus," *Annals of the New York Academy of Sciences* 521 (1988): 59–71. K. Bulloch and R. Y. Moore, "Innervation of the thymus gland by brain stem and spinal cord in mouse and rat," *American Journal of Anatomy* 162 (1981): 157–166.

DAVID L. FELTEN

D. L. Felten, S. Y. Felten, D. L. Bellinger, S. L. Carlson, K. D. Ackerman, K. S. Madden, J. A. Olschowski, and S. Livnat, "Noradrenergic sympathetic neural interactions with the immune system: Structure and function," *Immunological Reviews* 100 (1987): 225–260.

The following articles by Dr. David Felten and colleagues appear in both the *Journal of Neuroscience Research* 18 (1987) and in *Neuroimmunomodulation*, edited by J. R. Perez-Polo, K. Bulloch, R. H. Angeletti, G. A. Hashim, and J. de Vellis (New York: Alan R. Liss, 1987):

D. L. Felten, K. D. Ackerman, S. J. Wiegand, and S. Y. Felten, "Noradrenergic sympathetic innervation of the spleen: I. Nerve fibers associate with lymphocytes and macrophages in specific compartments of the splenic white pulp": 28–36. S. Y. Felten and J. Olschowka, "Noradrenergic sympathetic innervation of the spleen: II. Tyrosine hydroxylase (TH)-positive nerve terminals form synapticlike contacts on lymphocytes in the splenic white pulp": 37–48. K. D. Ackerman, S. Y. Felten, D. L. Bellinger, and D. L. Felten, "Noradrenergic sympathetic innervation of the spleen: III. Development of innervation in the rat spleen": 49–54. D. L. Bellinger, S. Y. Felten, T. J. Collier, and D. L. Felten, "Noradrenergic sympathetic innervation of the spleen: IV. Morphometric analysis in adult and aged F344 rats": 55–63. S. L. Carlson, D. L. Felten, S. Livnat, and S. Y. Felten, "Noradrenergic sympathetic innervation of the spleen: V. Acute drug-induced depletion of lymphocytes in the target fields of innervation results in redistribution of noradrenergic fibers but maintenance of compartmentation": 64–69.

BRANSISLAV D. JANKOVIĆ

B. D. Janković, "Neural tissue hypersensitivity in psychiatric disorders with immunologic features," *The Journal of Immunology* 135 (1985): 835–857. B. D. Janković, "Neuroimmune interactions: Experimental and clinical strategies," *Immunology Letters* 16 (1987): 341–354. B. D. Janković and K. Isaković, "Neuroendocrine correlates of immune response. I. Effects of brain lesions on antibody production, arthus reactivity and delayed hypersensitivity in the rat," *International Archives of Allergy and Applied Immunology* 45 (1973): 360–372.

MARK L. LAUDENSLAGER

M. L. Laudenslager, "The psychobiology of loss: Lessons from humans and non-human primates," *Journal of Social Issues* 44 (1988): 19–36. M. L. Laudenslager, S. M. Ryan, R. C. Drugan, R. L. Hyson, and S. F. Maier, "Coping and immunosuppression: Inescapable but not escapable shock suppresses lymphocyte proliferation," *Science* 221 (1983): 568–570. S. F. Maier and M. L. Laudenslager, "Inescapable shock, shock controllability, and mitogen-stimulated lymphocyte proliferation," *Brain, Behavior, and Immunity* 2 (1988): 87–91.

CANDACE B. PERT

C. B. Pert, J. M. Hill, M. R. Ruff, R. M. Berman, W. G. Robey, L. O. Arthur, F. W. Ruscetti, and W. L. Farrar, "Octapeptides deduced from the neuropeptide receptor-like pattern of antigen T4 in brain potentially inhibit human immunodeficiency virus receptor binding and T-cell infectivity," *Proceedings of the National Academy of Sciences* 83 (1986): 9254–9258

WALTER PIERPAOLI

W. Pierpaoli, H. G. Kopp, J. Mueller, and M. Keller, "Interdependence between neuroendocrine programming and the generation of immune recognition in ontogeny," *Cellular Immunology* 29 (1977): 16–27. W. Pierpaoli and G. J. M. Maestroni, "Melatonin: A principle neuroimmunoregulatory and anti-stress hormone: Its anti-aging effects," *Immunology Letters* 16 (1987): 355–362.

NOVERA HERBERT SPECTOR

Novera H. Spector, "Old and new strategies in the conditioning of immune responses," from the Proceedings of the Second International Workshop on Neuroimmunomodulation, *The Annals of the New York Academy of Sciences* 496 (June 1987): 522–531. V. Ghanta, R. Hiramoto, B. Solvason, and N. H. Spector, "Neural and environmental influences on neoplasia and conditioning of NK activity," *Journal of Immunology* 135 (1985): 848S–852S.

HERBERT WEINER

Herbert Weiner, *Psychobiology and Human Disease* (New York: American Elsevier, 1977). Herbert Weiner, "Special article: The dynamics of the organism: Implications of recent biological thought for psychosomatic theory and research," *Psychosomatic Medicine* (1989). Forthcoming.

23. A Portfolio of Related Matters

Honor Roll

More information about the individuals listed in the honor roll can be obtained from the following references:

FRANZ GABRIEL ALEXANDER

F. G. Alexander, *Psychosomatic Medicine* (New York: W. W. Norton, 1950).

ARISTOTLE

Aristotle, "De anima," in J. A. Smith and W. D. Ross, editors, *The Works of Aristotle*, vol. 3 (Oxford: Oxford University Press, 1931). Aristotle, *On Youth and Old Age, Life and Death, and Respiration*, translated by W. Ogle (London: Longmans, Green, 1897).

HENRY K. BEECHER

H. K. Beecher, "Evidence for increased effectiveness of placebos with increased stress," *American Journal of Physiology* 187 (1956): 163b. H. K. Beecher, "The powerful placebo," *Journal of the American Medical Association* 159 (1955): 1602–1606.

CLAUDE BERNARD

C. Bernard, *Leçons de Physiologie Expérimentale Appliquée à la Médicine*, 2 vols. (Paris: J. B. Ballière, 1855–56). J. M. D. Olmstead, *Claude Bernard, Physiologist* (New York: Harper and Brothers, 1938).

WALTER BRADFORD CANNON

W. Cannon, *Bodily Changes in Pain, Hunger, Fear and Rage* (New York: W. W. Norton, 1963). W. Cannon, "The James-Lange Theory of Emotions: A Critical Examination and an Alternative Theory," in Karl Pribram, editor, *Brain and Behavior*, vol. 4 (Harmondsworth, Eng.: Penguin Books, 1969), 433–451. W. Cannon, *The Wisdom of the Body* (New York: W. W. Norton, 1939).

JEAN MARTIN CHARCOT

J. M. Charcot, *Diseases of the Nervous System*, vol. 3 (London: Sydenham Society, 1889).

RENÉ JULES DUBOS

R. Dubos, *Man Adapting* (New Haven: Yale University Press, 1965). R. Dubos, *Man, Medicine and Environment* (New York: Praeger, 1968). R. Dubos, *Mirage of Health* (New York: Harper and Row, 1971. R. Dubos, "The state of health and the quality of life," *The Western Journal of Medicine* 125 (1976): 8–9.

HELEN FLANDERS DUNBAR

H. F. Dunbar, *Emotions and Bodily Changes* (New York: Columbia University Press, 1954).

JOEL ELKES

L. J. Dickstein and J. Elkes, "Health awareness and the medical student: A preliminary experiment," *Advances* 4 (1988): 11–23.

GEORGE L. ENGEL

G. L. Engel, "The need for a new medical model: A challenge for biomedicine," *Science,* 196 (1977): 129–136.

JEROME D. FRANK

J. Frank, *Persuasion and Healing* (New York: Schocken Books, 1961). J. Frank, "The faith that heals," *The Johns Hopkins Medical Journal,* 137 (1975): 127–131.

GALEN

The medical philosophy of Galen can be found in: L. Clendening, *Source Book of Medical History* (New York: Dover Publications, 1942), 41–51. *Galen: Three Treatises on the Nature of Science,* translated by Richard Walzer and Michael Frede (Indianapolis: Hackett, 1985), 55–56.

LAWRENCE J. HENDERSON

L. J. Henderson, *The Order of Nature* (Cambridge: Harvard University Press, 1925).

HIPPOCRATES

The medical philosophy of Hippocrates can be found in Clendening, *Source Book of Medical History,* 13–26. *The Genuine Works of Hippocrates,* translated by Francis Adams (Huntington, New York: Robert E. Krieger, 1972).

JIMMIE C. D. HOLLAND

M. A. Andrykowski, P. B. Jacobsen, E. Marks, K. Gorfinkle, T. B. Hakes, R. J. Kaufman, V. E. Currie, J. C. Holland, and W. H. Redd, "Prevalence, predictors and course of anticipatory nausea in women receiving adjuvant chemotherapy for breast cancer," *Cancer* 62 (1988): 2607–2613. W. Breitbart and J. C. Holland, "Psychiatric complications of cancer," in M. C. Brain and P. O. Carbone, editors, *Current Therapy in Hematology-Oncology-*3 (Burlington, Ontario, Canada: B. C. Decker, 1988), 268–274. P. B. Jacobsen, M. A. Andrykowski, W. H. Redd, M. Die-Trill, T. B. Hakes, R. J. Kaufman, V. E. Currie, and J. C. Holland, "Nonpharmacologic factors in the development of post-treatment nausea with adjuvant chemotherapy in breast cancer," *Cancer* 61 (1988): 379–385. J. C. Holland and S. Tross, "The psychological and neuropsychiatric sequelae of the Acquired Immunodeficiency Syndrome and related disorders," *Annals of Internal Medicine* 103 (1985): 760–764. M. J. Massie and J. C. Holland, "The cancer patient with pain: Psychiatric complications and their management," *Medical Clinics of North America* 71 (1987): 243–258.

OLIVER WENDELL HOLMES

O. W. Holmes, *Addresses and Exercises at the One Hundredth Anniversary of the Foundation of the Medical School of Harvard University, October 17, 1883* (Cambridge: John Wilson and Son, University Press, 1884), 17–18. O. W. Holmes, *Currents and Counter-Currents in Medical Science with Other Addresses and Essays* (Boston: James R. Osgood and Company, 1878). O. W. Holmes, *Collection of Medical Essays (*1842–1882) (Boston: Houghton Publishing, 1883). Contains 400-plus essays.

FRANZ J. INGELFINGER

F. J. Ingelfinger, "Arrogance," *New England Journal of Medicine* 303 (1980): 1506–1511. F. J. Ingelfinger, "Those anti-doctor books," *New England Journal of Medicine* 294 (1976): 442–443.

WILLIAM JAMES

W. James, *Psychology* (New York: World Publishing, 1948).

DAVID M. KISSEN

D. M. Kissen, "Psychosocial factors, personality and lung cancer in men aged 55–64," *British Journal of Medical Psychology* 40 (1967): 29–43.

LAWRENCE LeSHAN

L. LeShan, "A basic psychological orientation apparently associated with malignant disease," *The Psychiatric Quarterly* 36 (1961): 314–330. L. LeShan and R. E. Worthington, "Personality as a factor in the pathogenesis of cancer: A review of the literature," *British Journal of Medical Psychology* 29 (1956): 49–56. L. LeShan, *You Can Fight For Your Life: Emotional Factors in the Causation of Cancer* (New York: M. Evans, 1977).

MOSES MAIMONIDES

Maimonides' Medical Writings, translated and annotated by Fred Rosner (Haifa, Israel: The Maimonides Research Institute, 1987). *The Medical Aphorisms of Moses Maimonides*, translated and edited by

Fred Rosner and Suessman Muntner (New York: Yeshiva University Press, 1971), 41–53.

KARL MENNINGER

K. Menninger, with M. Mayman and P. Pruyser, *The Vital Balance: The Life Process in Mental Health and Illness* (New York: Viking, 1963).

NEAL E. MILLER

N. E. Miller, "Behavioral medicine: Symbiosis between laboratory and clinic," *Annual Review of Psychology* 34 (1983), 1–31. N. E. Miller and B. S. Brucker, "A learned visceral response apparently independent of skeletal ones in patients paralyzed by spinal lesions," in N. Birbaumer and H. D. Kimmel, editors, *Biofeedback and Self-Regulation* (Hillsdale, N.J.: Lawrence Erlbaum and Associates, 1979), 287–304.

SIR WILLIAM OSLER

Sir William Osler, *Aequanimitas*, 3rd edition (New York: Blakiston Company, 1953). W. Osler, "The faith that heals," *British Medical Journal* 1 (1910): 1470–1472. H. A. Christian, editor, *Osler's Principles and Practice of Medicine*, 15th edition (New York: D. Appleton-Century, 1944).

PARACELSUS

The medical philosophy of Paracelsus can be found in L. Clendening, *Source Book of Medical History* (New York, Dover Publications, 1942), 95–105.

IVAN PETROVICH PAVLOV

I. P. Pavlov, *Lectures on Conditioned Reflexes* (New York: Liveright, 1928). I. P. Pavlov, *Lectures on the Work of the Principal Digestive Glands* (St. Petersburg: I. N. Kushnereff, 1897).

KARL H. PRIBRAM

K. H. Pribram, *Brain and Perception* (Hillsdale, N.J.: Lawrence Erlbaum and Associates, 1990, forthcoming). K. H. Pribram, "The Four R's of Remembering," in K. H. Pribram, editor, *On the Biology of Learning* (New York: Harcourt, Brace and World, 1969), 193–225. K. H. Pribram, "The new neurology and the biology of emotion: A structural approach," in K. H. Pribram, editor, *Brain and Behavior*, vol. 4 (Harmondsworth, Eng.: Penguin Books, 1969), 452–468. K. H. Pribram, "Proposal for a structural pragmatism: Some neuropsychological considerations of problems in philosophy," in K. H. Pribram, editor, *Brain and Behavior*, vol. 1 (Harmondsworth, Eng.: Penguin Books, 1969), 11–19. K. H. Pribram, "Proposal for a structural pragmatism: Some neuropsychological considerations of problems in philosophy," in K. H. Pribram, editor, *Brain and Behavior*, vol. 4 (Harmondsworth, Eng.: Penguin Books, 1969), 494–505.

AARON FREDERICK RASMUSSEN, JR.

A. F. Rasmussen, Jr., "Emotions and immunity," *Annals of the New York Academy of Sciences* 164 (1969): 458–461.

VERNON S. RILEY

V. Riley, M. A. Fitzmaurice, and D. H. Spackman, "Psychoneuroimmunologic factors in neoplasia: Studies in animals," in R. Ader, editor, *Psychoneuroimmunology* (New York: Academic Press, 1981), 31–102.

JONAS EDWARD SALK

J. Salk, "The mind of man," in: J. Salk, editor, *Man Unfolding* (New York: Harper and Row, 1972), 107–112.

HANS SELYE

H. Selye, *The Stress of Life* (New York: McGraw-Hill, 1956).

BENEDICTUS DE SPINOZA

B. de Spinoza, "On the origin and nature of the emotions," in C. H. Bruder, editor, *The Ethics* (Leipzig: Tauchnitz, 1843).

WILLIAM HENRY WELCH

Walter Burkett, editor, *Papers and Addresses by William Henry Welch*, 3 vols. (Baltimore: The Johns Hopkins Press, 1920). Simon Flexner and James Flexner, *William Henry Welch and the Heroic Age of American Medicine* (New York: Viking Press, 1941).

STEWART WOLF

S. Wolf, "The pharmacology of placebos," *Pharmacological Reviews*, 11 (1959): 698–704. S. Wolf and R. H. Pinsky, "Effects of placebo administration and occurrence of toxic reactions," *Journal of the American Medical Association* 155 (1974): 339–341.

HAROLD G. WOLFF

H. G. Wolff, *Stress and Disease*, revised and edited by Stewart Wolf and Helen Goodell (Springfield, Ill.: Charles C. Thomas, 1968).

Ongoing Studies in Biochemical Pathways

A. Kling, K. H. Tachiki, A. Steinberg, P. Lucas, C. Kessler, N. Sachinvala, H. Von Scoti, M. Terpenning, P. Shapshack, and M. Cohen, "A psychoneuroimmunological study of an unusual family cohort of multiple paranoid schizophrenic siblings," *Neuropsychiatry, Neuropsychology, and Behavioral Neurology* 1 (1988): 191–215.

Doctors and Nutrition

Dr. Robert A. Good's studies of diet and impaired immunological function are reported in R. A. Good and N. K. Day, "The influence of nutrition on autoimmunities, cancer, heart disease, and other diseases of aging," in Robert J. Morin and Richard J. Bing (eds.), *Frontiers in Medicine: Implications for the Future* (New York: Human Sciences Press, 1985), 207–221. R. A. Goode, "Nutrition, Immunity, Aging, and Cancer," *Nutrition Reviews 46* (1988): 62–67.

Dr. Roy L. Walford's research on restricted food intake and life expectancy was reported in the following publications: Roy L. Walford, *The Maximum Life Span* (New York: Avon, 1984). Richard Weindruch and Roy L. Walford, *The Retardation of Aging and Disease by Dietary Restriction* (Springfield, Ill.: Charles C. Thomas, 1988). Roy L. Walford, *The 120 Year Diet Plan: How to Double Your Vital Years* (New York: Simon and Schuster, 1986).

Dr. C. Everett Koop's report on the role of nutrition in health and disease, entitled, *The Surgeon General's Report on Nutrition and Health: Summary and Recommendations* (DHHS publication #88-50211), was published by the U.S. Public Health Service, Office of the Surgeon General, Department of Health and Human Services, 1988.

The study on nutrition education in medical schools from which excerpts in this chapter have been drawn was conducted by the National Research Council Committee on Nutrition in Medical Education. Their report, *Nutrition Education in U.S. Medical Schools*, was published by National Academy Press, Washington, D.C., in 1985.

To The Graduates: A Physician's Credo

Norman Cousins delivered a commencement address in 1982 to the graduating class of the George Washington University Medical Center discussing the role of the physician, which later appeared in the *Journal of the American Medical Association* 248 (1982): 587–589.

SUGGESTED READING

I gratefully acknowledge the assistance of Drs. Joan Borysenko and George Solomon in the compilation of this list.

POPULAR LITERATURE

ACHTERBERG, JEANNE. *Imagery in Healing: Shamanism and Modern Medicine*. Boston: New Science Library, 1985.

BENSON, HERBERT. *The Mind/Body Effect*. New York: Simon and Schuster, 1979.

BENSON, HERBERT, with MIRIAM Z. KLIPPER. *The Relaxation Response*. New York: Avon Books, 1976.

BENSON, HERBERT, and WILLIAM PROCTOR. *Beyond the Relaxation Response*. New York: Berkley, 1985.

BORYSENKO, JOAN. *Minding the Body, Mending the Mind*. New York: Bantam Books, 1987.

CAPRA, FRITJOF. *The Tao of Physics*. New York: Bantam Books, 1977.

DOSSEY, LARRY. *Space, Time and Medicine*. Boston: Shambhala Publications, 1982.

DREHER, HENRY. *Your Defense Against Cancer*. New York: Harper and Row, 1988.

EISENBERG, DAVID, with THOMAS LEE WRIGHT. *Encounters with Qi: Exploring Chinese Medicine*. New York: W. W. Norton, 1985.

JAFFE, DENNIS. *Healing from Within*. New York: Simon and Schuster, 1980.

KORN, ERROL R., and KAREN JOHNSON. *Visualization: The Uses of Imagery in Health Professions*. Homewood, Illinois: Dow-Jones-Irwin, 1983.

LOCKE, STEVEN, and DOUGLAS COLLIGAN. *The Healer Within: The New Medicine of Mind and Body*. New York: E. P. Dutton, 1986.

MATTHEWS-SIMONTON, STEPHANIE, O. CARL SIMONTON, and JAMES L. CREIGHTON. *Getting Well Again*. New York: Bantam Books, 1978.

PELLETIER, KENNETH. *Mind as Healer, Mind as Slayer*. New York: Dell Books, 1977.

PELLETIER, KENNETH. *Longevity: Fulfilling Our Biological Potential*. New York: Dell Books, 1982.

SELIGMAN, MARTIN E. P. *Helplessness: On Depression, Development and Death*. New York: W. H. Freeman, 1975.

SIEGEL, BERNIE S. *Love, Medicine and Miracles*. New York: Harper and Row, 1986.

WEIL, ANDREW. *Health and Healing: Understanding Conventional and Alternative Medicine*. Boston: Houghton Mifflin, 1983.

MORE TECHNICAL LITERATURE

ADER, ROBERT, ed. *Psychoneuroimmunology*. New York: Academic Press, 1981.

BERGSMA, DANIEL, and ALLAN L. GOLDSTEIN, eds. *Neurochemical and Immunologic Components in Schizophrenia*. Proceedings of a conference held at the University of Texas Medical Branch, October 1976. New York: Alan R. Liss, 1978.

BRIDGE, T. PETER, ALLAN F. MIRSKY, and FREDERICK K. GOODWIN, eds. *Psychological, Neuropsychiatric and Substance Abuse Aspects of AIDS*. New York: Raven Press, 1988.

COOPER, EDWIN L., ed. *Stress, Immunity and Aging*. New York: Marcel Dekker, 1984.

COTMAN, CARL W., ROBERTA E. BRINTON, ALBERT GALABURDA, BRUCE MCEWEN, and DIANA M. SCHNEIDER, eds. *The Neuroimmune Endocrine Connection*. New York: Raven Press, 1987.

DILMAN, VLADIMIR M. *The Law of Deviation of Homeostasis and Diseases of Aging*. Boston: John Wright PSG/ 1981.

FREDERICKSON, ROBERT C. A., HUGH C. HENDRIE, JOSEPH N. HINGTGEN, and MORRIS H. APRISON, eds. *Neuroregulation of Autonomic, Endocrine and Immune Systems*. Boston: Martinus Nijhoff, 1986.

GOETZL, EDWARD J., ed. Proceedings of a conference on Neuromodulation of Immunity and Hypersensitivity, Coconut Grove, Florida, November 1984. *The Journal of Immunology* 135 (supplement) (1985).

GUILLEMIN, ROGER, MELVIN COHN, and THEODORE MELNECHUK, eds. *Neural Modulation of Immunity*. Proceedings of an international symposium held under the auspices of the Princess Liliane Cardiology Foundation in Brussels, Belgium, October, 1983. New York: Raven Press, 1985.

JANKOVIĆ, BRANISLAV D., BRANISLAV M. MARKOVIC, and NOVERA HERBERT SPECTOR, eds. *Neuroimmune Interactions*. Proceedings of the Second International Workshop on Neuroimmunomodulation, Dubrovnik, Yugoslavia, June 1986. *Annals of the New York Academy of Sciences* 521 (1988).

KORNEVA, ELENA A., VIKTOR M. KLIMENKO, and ELENORA K. SHKHINEK. *Neurohumoral Maintenance of Immune Homeostasis*, translated from the Russian by Samuel A. and Elizabeth O'Leary Corson in collaboration with Roland Dartau, Justina Epp, and L. A. Mutschler. Chicago: University of Chicago Press, 1985.

LEVY, SANDRA M., ed. *Biological Mediators of Behavior and Disease: Neoplasia*. Proceedings of a symposium on Behavioral Biology and Cancer held at the National Institutes of Health, Bethesda, Maryland, May 1981. New York: Elsevier, 1982.

LLOYD, RUTH, ed., and GEORGE F. SOLOMON, consulting ed., and Martin E. Dorf, consultant in immunology. *Explorations in Psychoneuroimmunology*. Orlando, Florida: Grune and Stratton, 1987.

LOCKE, STEVEN E., ed. *Psychological and Behavioral Treatments for Disorders Associated with the Immune System: An Annotated Bibliography*. New York: Institute for the Advancement of Health, 1986.

LOCKE, STEVEN, ROBERT ADER, HUGO BESEDOVSKY, NICHOLAS HALL, GEORGE SOLOMON, and TERRY STROM, eds., and N. HERBERT SPECTOR, consulting ed., *Foundations of Psychoneuroimmunology*. New York: Aldine, 1985.

LOCKE, STEVEN E., and MADY HORNIG-ROHAN, eds. *Mind and Immunity: Behavioral Immunology*. New York: Institute for the Advancement of Health, 1983. Contains 1,453 research abstracts and references.

PEREZ-POLO, J. REGINO, KAREN BULLOCH, RUTH HOGUE ANGELETTI, GEORGE A. HASHIM, and JEAN DE VELLIS, eds., *Neuroimmunomodulation*. New York: Alan R. Liss, 1987.

PIERPAOLI, WALTER, and NOVERA HERBERT SPECTOR, eds. *Neuroimmunomodulation: Interventions in Aging and Cancer*. Proceedings of the First Stromboli Conference on Aging and Cancer, Stromboli, Sicily, June 1987. *Annals of the New York Academy of Sciences* 496 (1987).

PLOTNIKOFF, NICHOLAS P., ROBERT E. FAITH, ANTHONY J. MURGO, and

ROBERT A. GOOD, eds. *Enkephalins and Endorphins: Stress and the Immune System.* New York: Plenum Press, 1986.

SPECTOR, NOVERA HERBERT, ed.-in-chief. *Neuroimmunomodulation.* Proceedings of the First International Workshop on Neuroimmunomodulation, Washington, D. C., November 1984. New York: Gordon and Breach, 1988.

TEMOSHOK, LYDIA, guest ed. Special Issue on Acquired Immune Deficiency Syndrome (AIDS), *Journal of Applied Social Psychology* 17(3) (1987).

TEMOSHOK, LYDIA, CRAIG VAN DYKE, and LEONARD S. ZEGANS, eds. *Emotions in Health and Illness.* New York: Grune and Stratton, 1983.

SUMMARIAL ARTICLES

GORMAN, JAMES R., and STEVEN E. LOCKE. "Neural, endocrine and immune interactions." In *Comprehensive Textbook of Psychiatry/V*, edited by Harold I. Kaplan and Benjamin J. Sadock. Baltimore: Williams and Wilkins, 1989.

LOCKE, STEVEN E., and JAMES R. GORMAN. "Behavior and immunity." In *Comprehensive Textbook of Psychiatry/V*, edited by Harold I. Kaplan and Benjamin J. Sadock. Baltimore, Maryland: Williams and Wilkins, 1989.

MELNECHUK, THEODORE. "Emotions, brain, immunity, and health: A review." In *Emotions and Psychopathology*, edited by M. Clynes and J. Panksepp. New York: Plenum Press, 1988, 181–247. Contains 530 references.

PELLETIER, KENNETH R., and DENISE L. HERZING. "Psychoneuroimmunology: Toward a Mind-Body Model." *Advances* 5 (1) (1988): 27–56.

SOLOMON, GEORGE F. Psychoneuroimmunology: Interactions between central nervous system and immune system." *Journal of Neuroscience Research* 18(1) (1987): 1–9. This article also appears in *Neuroimmunomodulation* edited by J. R. Perez-Polo, K. Bulloch, R. H. Angeletti, G. A. Hashim, and J. de Vellis. New York: Alan R. Liss, 1987, 1–9.

JOURNALS

Advances, published by the Institute for the Advancement of Health, 16 East 53rd Street, New York, NY 10022.

Brain, Behavior, and Immunity, edited by Robert Ader, published by Academic Press, Inc., 1 East First Street, Duluth, Minnesota 55802.

INDEX